Transnational Environmental Policy

'This readable book is the best treatment of the subject published so far'

Jim Lovelock, Honorary Visiting Fellow, Green College, Oxford

'Stimulating and thought-provoking'

F. Sherwood Rowland, University of California, Irvine

Transnational Environmental Policy analyses a surprising success story in the field of international environmental policy making: the threat to the ozone layer posed by industrial chemicals and how it has been averted. The book also raises the more general question about problem-solving capacities of industrialised countries and the world society as a whole.

This case study investigates the regulations that have been put in place at an international level, and how the process evolved over twenty years in the USA and Germany. At the same time, it highlights problem-solving capacities of industrialised countries: is the international community in a position to tackle global environmental threats? Under which conditions is transnational governance without government possible?

Combining insights from political science and sociology, Reiner Grundmann develops a policy network approach that traces environmental advocacy in transnational settings. He analyses key scientific controversies based on insights in the sociology of science; and examines risk sociology, institutional analysis and cultural theory in order to understand the role of discourses, norms and ideas in decisions under uncertainty. Based on expert interviews, archive material and mass media analysis, this book challenges commonly accepted accounts of the case which have so far been put forward. Finally, Grundmann suggests lessons to be learnt from the ozone layer scenario and applies them to the case of global climate change.

This fascinating study will be invaluable for students and researchers in the sociology of science, public policy and regulation, global environmental and heath problems; and environmental sociology.

Reiner Grundmann is Senior Lecturer at the Aston Business School. He is a sociologist and political scientist. He is the author of *Marxism and Ecology*, and recently co-edited essays by Werner Sombart.

Routledge Studies in Science, Technology and Society

Transnational Environmental Policy

Reconstructing ozone

Reiner Grundmann

Routledge
Taylor & Francis Group

LONDON AND NEW YORK

First published 2001
by Routledge

2 Park Square, Milton Park, Abingdon, Oxfordshire OX14 4RN
52 Vanderbilt Avenue, New York, NY 10017

Routledge is an imprint of the Taylor & Francis Group, an informa business

First issued in paperback 2020

Typeset in Baskerville by RefineCatch Ltd

British Library Cataloguing in Publication Data
A catalogue record for this book is available from the British Library

Library of Congress Cataloging in Publication Data
Grundmann, Reiner.
   Transnational environmental policy: reconstructing ozone / Reiner Grundmann.
       p.   cm. – (Routledge studies in science, technology, and society; 3)
   Includes bibliographical references and index.
   ISBN 0–415–22423–3 (alk. paper)
    1. Ozone layer depletion.   2. Environmental policy–International cooperation.   I. Title.   II. Series.
QC879.7 .G78 2001
363.738′75′056–dc21   00–062789

ISBN 978-0-415-22423-9 (hbk)
ISBN 978-0-367-60489-9 (pbk)

To Gabriella, Leon and Rebecca

Contents

Tables

Figures

Preface and acknowledgements

This book is based on my previous book *Transnationale Umweltpolitik*, which was published in 1999 by Campus. However, it is not a true translation since I have had the opportunity to make several changes. Compared to the German version, I have tried to present the main points in a more convenient way for a different readership, which basically means I have recast the social science discussion by using approaches published in English.

I have rearranged and shortened some chapters, but also elaborated the argument at times, above all in the first chapter, which now contains a discussion of cultural theory and social constructivist approaches. I added a new Chapter 8, the epilogue, where I try to interpret the case of global climate change along the lines of the approach taken in this book.

Jim Lovelock and Sherry Rowland both read the whole manuscript and made valuable comments. Their thorough reading and detailed comments helped increase the accuracy of my presentation of arguments from the scientific controversy. It goes without saying that I am extremely grateful to them. Both have been protagonists of the controversy, so it was crucial to get their feedback.

I would also like to thank Paul M. Malone who translated large parts of the manuscript. All quotes from German sources have been translated into English.

Last but not least, my thanks go to my family who competed with me for access to the home computer. A computer literate family is a mixed blessing. On the one hand, they do not perceive me as an extension of the PC (a 'nerd'); the drawback is that they also demand computing time. We managed to reach agreements mostly by striking simple bargains. When demands became passionate, we had to resort to more comprehensive solutions. This seems to be congenial to topics discussed in this book: management of a common pool resource and variants of problem-solving. We could, in principle, have solved this problem by purchasing more computers, ideally one for every family member. This is where the analogy to the ozone case breaks down: there is only one ozone layer.

Birmingham, UK
July 2000

Abbreviations

AAOE	Airborne Antarctic Ozone Experiment
AFEAS	Alternative Fluorocarbons Acceptability Study
AgV	Arbeitsgemeinschaft der Verbraucher
AKE	Arbeitskreis Energie der Deutschen Physikalischen Gesellschaft
ANPR	Advance Note of Proposed Rulemaking
BACER	Biological and Climatic Effects Research
BAS	British Antarctic Survey
BMFT	Bundesministerium für Forschung und Technologie
BMI	Bundesministerium des Innern
BMU	Bundesministerium für Umwelt, Naturschutz und Reaktorsicherheit
BMW	Bundesministerium für Wirtschaft
CCOL	Coordinating Committee on the Ozone Layer
CEQ	Council on Environmental Quality
CFCs	Chlorofluorocarbons
CIAP	Climatic Impact Assessment Program
ClO	Chlorine monoxide
CLP	Chlorine Loading Potential
CMA	Chemical Manufacturers Association
COAS	Council on Atmospheric Sciences
CPR	Common Pool Resources
CPSC	Consumer Product and Safety Commission
DFG	Deutsche Forschungsmeinschaft
DMG	Deutsche Meteorologische Gesellschaft
DOC	Department of Commerce
DOT	Department of Transportation
DPG	Deutsche Physikalische Gesellschaft
EC	European Community
EFCTC	European Fluorocarbon Technical Panel
EK	Enquetekommission 'Vorsorge zum Schutz der Erdatmosphäre' des Deutschen Bundestages
EPA	Environmental Protection Agency
FCCC	Framework Convention on Climate Change
FDA	Food and Drug Administration

GEF	Global Environmental Facility
GWP	Global Warming Potential, Treibhauspotential
HCFCs	Hydrochlorofluorocarbons
HFCs	Hydrofluorocarbons
ICI	Imperial Chemical Industries
IGA	Industriegemeinschaft Aerosole
IMOS	(Ad hoc Federal Interagency Task Force on the) Inadvertent Modification of the Stratosphere
IPCC	Intergovernmental Panel on Climate Change
JPL	Jet Propulsion Laboratory
MCA	Manufacturing Chemists Association, later: CMA
MLF	Multilateral Fund
NAS	National Academy of Sciences
NASA	National Aeronautics and Space Administration
NCAR	National Centre for Atmospheric Research
NGOs	Non-governmental Organizations
NOAA	National Oceanic and Atmospheric Administration
NOZE	National Ozone Expedition
NRDC	Natural Resources Defence Council
NSF	National Science Foundation
ODP	Ozone Depletion Potential
ODS	Ozone Depleting Substances
OTP	Ozone Trends Panel
SCI	Science Citation Index
SRU	Sachverständigenrat für Umweltfragen
SSK	Sociology of Scientific Knowledge
TOMS	Total Ozone Mapping Spectrometer
UBA	Umweltbundesamt
UNDP	United Nations Development Program
UNEP	United Nations Environmental Program
VCI	Verband der chemischen Industrie
WMO	World Meteorological Organization

1 Social science and global environmental problems

Every year in the autumn there are fresh news reports on the ozone hole over the Antarctic. There is an almost ritual quality to the coverage, since the script always seems to be the same: in comparison with the preceding year, the hole in the ozone layer has grown or the ozone levels have reached new record lows. Obviously, only negative records are set in this area. The lay public is given the impression that too little is being done in the face of this catastrophic development. In actuality, the international community has already agreed upon measures that could solve the problem. According to the experts, however, it will be several more decades before the seasonally appearing ozone hole over the South Pole disappears (WMO 1994). This is due above all to the longevity of the ozone-destroying substances, which remain in the atmosphere for a long period after their release. Thus far the facts – paradoxical as they may seem at first – are simple: although the ozone layer initially continues to deteriorate, in the long term the measures are assumed to be successful. The matter becomes more complicated, however, if we want to understand how it became possible to arrive at binding controls in the first place.

This case study analyses a surprising success story in the field of international environmental policy making. It investigates the regulations that have been put in place at the international level and how the process evolved over twenty years in the USA and Germany. This raises the more general question about problem-solving capacities of industrialised countries. Is the international community in a position to tackle global environmental threats? Under which conditions is transnational governance without government possible? (Rosenau and Czempiel 1992).[1]

Global ecological problems are newcomers to politics and a new research topic in the social sciences. To be sure, the 1960s saw international agreements to ban nuclear tests in the atmosphere, and back in the nineteenth century cholera and yellow fever have been the object of such attempts (Cooper 1989). Also in the 1960s, the first alarming reports about environmental problems were published. But it was not until the 1970s, with the publication of the reports of the Club of Rome, the establishment of the United Nations Environmental Programme (UNEP), and the publication of *Global 2000* and the Brundtland Report (World Commission 1987, making the term 'sustainable development' current) that these issues were established on the political agenda on a permanent basis. Since then, many problems have been thematised, ranging from the civil application of nuclear energy to genetic engineering and to global warming. To date, more than 170 multilateral agreements on the environment have been reached (UNEP 1993). Social scientists have also become aware of the problem. The exponential growth in literature testifies to this (among many: Beck 1992; Beck *et al.* 1995; Breitmeier 1996; Dickens 1996; Dryzek 1997; Gehring 1994; Haas *et al.* 1993; Irwin 1995; MacNaghten and Urry 1998; Martell 1994; Murphy 1997; Norgaard 1994; Redclift and Benton 1994; Redclift and Woodgate 1996; Shiva 1989; Yearley 1996; Young 1994; Young and Osherenko 1993). However, it should be noted that interest from social science has been highly selective: 'Sociologists and anthropologists have concentrated on working scientists, whereas the few political scientists and policy analysts interested in science have focused on politicians' (Hart and Victor 1993: 643). Therefore, the challenge is to combine approaches from different social science backgrounds. I shall develop an institutionalist approach that incorporates insights from different areas of political science and sociology.

A success story

This book is about controversies. It is about the claims and counterclaims of scientists, industry, politicians and environmentalists. These controversies had to do with the ozone layer, seen as endangered by some, while others thought that no big threat was looming. In order to contextualise these disputes, let me start with some remarks about the object of controversy.

The ozone layer is located approximately between 10 and 50 kilometres above sea-level and provides protection from certain injurious wavelengths of ultraviolet (UV) sunlight, above all from UV-B. Any reduction of the ozone column leads to an increase in the UV-B radiation reaching the earth. Before chlorofluorocarbons (CFCs) were seen as possible hazards to the ozone layer, suspicion fell in particular on nuclear explosions and aeroplane exhaust gases. The latter were examined by the Climatic Impact Assessment Program from 1971 to 1973, a US government technology assessment of possible effects of a planned fleet of supersonic aircraft (for example, of the Concorde type). There were fears that the operation of 500 supersonic aircraft could deplete the ozone layer by 22 per cent (Johnston 1971). A later study demonstrated that the risks were considerably lower (CIAP 1973). At about the same time, scientists began to suspect that chlorine from the exhaust of planned space shuttles could have a much greater effect on the ozone layer (Stolarski and Cicerone 1974). As a result, researchers became aware for the first time of the possibility that chlorine poses a potential danger to the ozone layer. In the same year, chlorofluorocarbons were singled out as possible ozone destroyers (Molina and Rowland 1974).

In 1974, scientists advanced the hypothesis that CFCs could damage the ozone layer. By way of increased UV-B radiation, severe environmental and health problems would follow. In the USA, this hypothesis was taken seriously. As a consequence, the Clean Air Act of 1977 introduced a CFC ban in spray-cans. Several other countries followed suit, but not the other large producers of CFCs in Japan or Europe. They pointed to the unknowns of the causal relationships. International negotiations saw two camps opposed to each other: the USA (and few other countries) who proposed strict emission controls for CFCs (despite scientific uncertainty) and the rest of the world. After a protracted negotiation period, a compromise was reached in September 1987 (Montreal Protocol) which envisaged a 50 per cent cut of ozone depleting substances by the year 1999. In the years that followed, this goal was made even more ambitious, resulting in a complete phaseout of these substances in the industrialised countries by 1996 (Benedick 1991; Brack 1996; UNEP 1995). These agreements comprise binding control measures to curb CFC emissions. They are – together with subsequent amendments – an instance of successful international governance. According to the United Nations Environment Program, world production of CFCs has been halved in the period from 1986 to 1992 (UNEP 1995: 32).[2] This success was unexpected at the time and today still poses problems in terms of theoretical explanation.

The Montreal Protocol is *the* foremost example of global environmental cooperation. It limits the emissions of major ozone depleting substances, mainly CFCs and halons. International regulations for the protection of the ozone layer seem to have been successful.[3] They are even deemed to be a blueprint for other global environmental problems like climate change. One of the architects of the Montreal Protocol, former UNEP director Mostafa Tolba, is not alone in espousing such 'policy learning' when he says: 'The mechanisms we design for the protocol will – very likely – become the blueprint for the institutional apparatus designed to control greenhouse gases and adaptation to climate change' (cit. in Benedick 1991: 7; see also Downie 1995). I shall return to this question in the final chapter.

The theoretical problem

And yet, there is a puzzle. Four characteristics of the case seem to make its success unlikely. First, the successful protection of an environmental asset is unlikely since it has no strong lobby but strong enemies (Schnaiberg *et al.* 1986). Second, in contradistinction to regulations within the nation-state where the 'shadow of hierarchy' sets more favourable conditions for cooperation, regulations in the international arena are bound to take place in the 'shadow of anarchy' (Oye 1986). A third puzzle pertains to the character of the good that is at stake. The ozone layer is a common pool resource, not a public good. This distinction may be subtle, yet it has important consequences. A public good can (in principle) be provided by a single actor without the free-riding of others affecting it adversely. In contrast, unilateral action cannot produce or protect a common pool resource, but can harm or destroy it. As a consequence, all potential polluters of the atmosphere must be part of an international agreement to protect the ozone layer (i.e. the k-group is large).[4] Even more surprising is the successful cooperation if one considers, fourth, that political decisions have had to be taken under uncertainty. This is to say that although expert knowledge is moving to centre stage, it cannot give a clear and unambiguous judgement since we typically have (at least) two opposing expert views on the matter (Adams 1995; Hajer 1995; Jasanoff 1992, 1995; Jasanoff and Wynne 1998).

All four reasons seem to pose considerable difficulties for international cooperation. Therefore, its success seems quite remarkable. Several approaches have been applied to this case (see Haas 1992b, 1993; Litfin 1994; Maxwell and Weiner 1993; Oye and Maxwell 1994; Rowlands 1995; Sebenius 1992; Sprinz and Vaahtoranta 1994). One can distinguish between economic and cognitive explanations. The economic explanation holds that the influence of industry was all decisive to the outcome. Specifically, the American chemical firm Du Pont is said to have gained a technological advantage over its competitors in developing alternative substances. This, it is argued, put the firm in a position to actually favour international regulations of CFCs. The cognitive approach has scientific consensus as the explaining variable. According to this view, expert communities reached a common understanding of causal relationships that could swiftly be transformed into political regulations. However, closer examination reveals that no evidence can be found for the claim that Du Pont had a technological advantage over its competitors. Neither was a scientific understanding of ozone depletion available at the time of the signing of the Montreal Protocol.

The problem has both a national and an international dimension. In the mid-1970s, political regulations took place on a national level. At the beginning of the 1980s, the process shifted to international level and found its conclusion in international treaties that became binding in the jurisdiction of the signatory countries. On both levels there are serious questions about costs and benefits and the urgency to act. Thus, we have two questions to answer. First, why was there international cooperation? And, second, why was the reaction different with respect to two industrialised countries (Germany and the USA), both heavy producers and consumers of ozone depleting substances? Both countries were chosen because of their crucial role in this controversy. There were other countries who played an important role (like the UK, for example) and it would have been ideal to include these, too. However, there are limits as to what one researcher can do in a given time period.

This chapter falls into three parts. First, I develop the aspects that made this success so unlikely: the representation of diffuse interests, the problem of international cooperation, the problem of public goods and decisions under uncertainty. Then I shall introduce the theoretical approach and finally make some remarks on the method chosen.

The representation of diffuse interests

One of the most common obstacles to environmental protection is the asymmetry of costs and benefits. Pollution abatement benefits the majority of the population but confers costs upon the polluters. Based on this observation, it has been argued that although environmental concerns do not have a lobby, they do have strong political enemies (Heidenheimer *et al.* 1990; Schnaiberg *et al.* 1986). In order to gain a better understanding of the power of the polluters and the relative weakness of environmental interests, I shall draw upon a typology offered by Wilson (1980). Using two dichotomous dimensions, he classified costs and benefits by whether they are concentrated on a few or spread more widely. This yields four possible cases, to which four types of regulatory policy can be assigned:

1   Where costs and benefits are broadly spread, a *majority policy* can be expected; however, this will not always produce successful regulation. For this to come about, the matter must reach the political agenda, and the legitimacy and effectiveness of the proposed measures must be beyond dispute.
2   If the benefits are broadly spread and the costs concentrated on a small group, we can expect a form of policy regulation dependent on the activities of public interest groups or policy entrepreneurs, representatives or advocates, as there will be a significant incentive among those bearing the costs (who, moreover, are few and therefore easy to organise) to resist the imposition of the measures proposed, but hardly any incentive among the beneficiaries to actively pursue their interests. Environmental and health matters mostly fall into this category.
3   If the benefits are concentrated and the costs are broadly spread, this will very probably result in *clientelism*, as a few small and easy-to-organise groups will benefit and will thus have a great incentive to organise and mobilise. The costs are so low and so thinly spread that there is little incentive for a (diffuse) opposition to mobilise.
4   If both the benefits and the costs are concentrated on a few, *interest groups* will form around their special interests. In this case, subsidies will be paid or other advantages awarded to one small social group at the cost of another small group, and the wider public will have no interest in the outcome of this conflict.

In cases like that under discussion, the problem is that of the representation of diffuse interests (case 2 above), which calls for speakers or representatives. If these are not available, we would expect the well-organised interests to prevail. The representation can be carried out in various ways, such as by individual actors, by public interest groups (environmental pressure groups, citizens' associations, political parties) or by networks including actors from various organisations and institutions. An individual activist will probably appeal to the public and work at a judicial level. Political entrepreneurs can combine a number of relevant subjects and processes (policy streams, cf. Kingdon 1984). They will often prepare solutions before a problem becomes acute, using windows of opportunity. In addition, organisations and social movements will exert influence on the political decision-making process: environmental groups typically do this by mobilising those affected or through symbolic protests. Networks may also operate in these three dimensions; their chances of success will be greater if they include representatives of various organisations in a position to mobilise extensive resources, alongside individual actors. Networks may therefore provide a coordinating function across society, which may be greater than could be created by power or hierarchy (Mayntz 1993). If, for example, scientists emerge as speakers at the same time as political entrepreneurs take up the question with the support of social movements, there will be an effect of mutual reinforcement.

Governments across the globe realise that they cannot control all the factors that influence the environmental quality within their jurisdiction. Pollutants cross national boundaries easily, a fact that has been dramatically highlighted after the explosion of the nuclear plant at Chernobyl when radioactive clouds meandered over Europe. Effective environmental protection therefore needs international cooperation. Again, the problem of diffuse interest representation surfaces. The representation of diffuse interests at an international level can be carried out in a number of ways: by speakers and activists who enjoy worldwide visibility, by the policies of national states, or by international organisations.

International cooperation

Political scientists or economists not very familiar with this case would probably explain the international cooperation by invoking the mechanism of the prisoners' dilemma and its solution. In international relations literature, inter-national cooperation is regarded as likely if a limited number of key actors engage in repeated prisoner's dilemmas. A well-known approach assumes that actors in such constellations would be able to reach cooperation if one actor makes the first cooperative move, which is then reciprocated by the others. The shadow of the future provides that the actors continue cooperating through this kind of insur-ance mechanism (Axelrod 1984; Taylor 1987). As neo-institutionalist approaches have pointed out, such iterated prisoner's dilemmas can lead to institutionalised international regimes.[5] Both possibilities have to be excluded as an explanation for the case under consideration, which did not follow the logic of a prisoners' dilemma. At the beginning of the international negotiations in the 1980s, main parties to a prospective treaty found themselves in a deadlock.[6] This situation was transformed when the pro-regulation camp was no longer content with the status quo (which did not provide regulations). However, the camp opposing regulations continued to do so even as others were going to take action. Nor can the existence of an international regime be invoked to explain the solution of this collective action problem, since the regime was the result of a prior successful international cooperation (i.e. the Montreal Protocol).

The model of the prisoners' dilemma is not always helpful in understanding problems in international relations. In fact, CFC-producing countries had defined their preferences for a long time *unilaterally*, i.e. knowing well what the preferences of the others were. Thus, an important ingredient of the prisoner's dilemma model is absent: the uncertainty about the behaviour of the other players. The constellation we deal with is best described as *deadlock*. This deadlock was not broken by an insurance mechanism but by isolating the draggers and the rise to hegemony of those countries that aspired to a comprehensive problem-solving strategy.[7]

Compared to the local or national level, the number and heterogeneity of players in international negotiations can cause additional problems, but also additional means to resolve them. An asymmetric distribution of capabilities and preferences is seen as beneficial for cooperation (Keohane and Ostrom 1994). Actors pre-committing themselves and taking on a leadership role can reduce the uncertainty and complexity of options dramatically.

Public goods and common pool resources

Another spontaneous reaction of social scientists confronted with this case might be to apply the logic of public goods to the ozone layer. However, there are various problems associated with this concept. Following Olson (1965), there have been several attempts to classify public goods, all trying to connect the type of goods to the problem of cooperation. It has been noted that the usual distinction between private and public goods (as in Musgrave and Musgrave 1989) gives us two ill-defined mixed categories ('mixed goods', cf. Cornes and Sandler 1994; Malkin and Wildavsky 1991; Willke 1995). The first mixed case arises where there is no rivalry regarding usage and access can be limited; the second is where rivalry about usage exists but access is not limited. These mixed cases can be defined as 'club goods' and 'common pool resources', respectively (see Table 1.1).

Apart from clarifying the problem of mixed cases, this typology highlights the distinction between public goods and common pool resources (CPR). Public goods are used in common and there are no limits to access. Common pool resources are also freely accessible but there is rivalry about their use. From this ensues an under-provision of the goods (this is the classical collective action problem). In cases where the goods have already been produced (or already exist as a gift of nature), the market gives too few incentives to protect it. Non-exclusiveness and usage rivalry may give some actors short-term benefits through over-consumption of the goods. Here is the difference between common pool

Table 1.1 Typology of goods

		Rivalry of use	
		no	yes
Access	exclusive	Club goods	Private goods
	not exclusive	Public goods	Common pool resources

Source: after Snidal (1979).

resources and public goods: one actor's use of resources can cause negative consequences for all others. Every inhabitant of the earth profits from an inviolate ozone layer. The ozone layer keeps intact even when the world population increases (*ceteris paribus*). However, it may be harmed if one single actor emits large amounts of ozone depleting substances into the atmosphere. It follows that the protection of common pool resources is much more difficult than the provision of a public good that in principle can be provided by one actor.

Decisions under uncertainty

Since the ozone case exemplifies a risk debate, might not the risk calculus provide some guidance for a decision? The classical risk calculus (size of potential damage multiplied by its probability to happen, cf. Starr 1969) gives a clear criterion for action, especially where problems are well defined (Otway 1985).[8] Insurance companies take their cue from this and cost-benefit analysis relies on it. However, decisions under uncertainty cannot take a lead from this since there is no commonly accepted definition of the situation and no sufficiently large class to make comparisons with (Adams 1995; Elster 1979; Knight 1921; Marcus 1988; Schon 1982; Wynne 1992). This is to say that whenever size and urgency of an issue are unknown, we are dealing with decisions under uncertainty. Playing a game against nature might suggest that scientists have an important part in reducing uncertainties. Typically, however, in many instances they are not able to provide a commonly accepted definition of a situation. In the case of atmospheric change we cannot draw upon a class of comparable cases: we only have one planet earth. In such unique situations atmospheric scientists confront similar problems to social scientists. Neither can test their hypotheses experimentally, and therefore have to define the situation from theories and existing data which are partly historical.[9] Model calculations and scenarios predict more or less plausible future system states. If these calculations remain stable and are accepted by relevant modellers, we would in fact have a common definition of the situation. Things are different where predictions vary from year to year because model parameters change or because different modellers get differing results, or because experimental measurements contradict each other or model predictions. In such cases uncertainty comes to the fore and poses the thorny question, time and again, whether the varying predictions are a reason to act or to wait and see. The wait-and-see position suggests that further research into the problem may eventually create certainty.

The point can be made in a slightly different way by pointing to the role of information. Usually, it seems to be accepted that we will be able to make better decisions about risks if we have more information. Economists particularly seem to operate under this premise. However, multiple and conflicting risk reports will make decisions difficult for anyone facing them (Viscusi 1997). Experimental research suggests that people tend to place a greater weight on worst case scenarios and thus behave in a risk averse fashion (Fischhoff *et al.* 1981). This highlights the importance of how a risk issue is presented to the public, i.e. the inclusion of worst case scenarios is clearly against the interest of those who believe there is no great risk. It would seem that they want to limit the use of such scenarios and promote consensus risk estimates that focus on the mean and exclude the range of risks. But the crux, of course, is that we usually do not know what the mean risk is. So the question is: how do we select risks, and how do we rank them? Before we go on to address these questions, let us ponder the ideal of 'mean risks'.

As indicated, probabilistic estimates about risks rest on statistical inferences based on a large number of incidences, like injuries in road traffic compared to distances travelled. This method seems deeply ingrained in experts' minds. They tend to translate every decision under uncertainty into a risk calculus. However, there is a real problem with this approach. Because it seems so scientific, it lends itself especially well to a discourse based on experts' opinion, thereby not only contradicting a (presumably) risk averse public but silencing it. This is an important element of the technocratic model of policy making (or 'science-based' policy making as it is usually called in the UK, see Hajer 1993). Its central elements are that there is an objective measure of risk that can be established with some confidence; that only scientific experts are knowledgeable about it; that lay perceptions are different from this expert view but false; and finally, that therefore, the public needs to be educated (for a critical view, see Irwin 1995).

There are, however, variations across countries. National policy styles are routinised or institutionalised methods of dealing with issues, 'legitimate ways of doing things'. The technocratic model seems to have been adopted above all in consensual political systems of Europe with expert committees meeting behind closed doors and giving advice to governments. On the other hand, the USA follows an adversarial approach where no culture of consensus opinion exists. Instead, extremes clash in public. The American institutional design allows for worst case scenarios, and thus for public involvement in an adversarial process. However, there may be different variants within each approach, for example, a more industry-friendly or more environment-friendly policy. Such differences in institutional structure can influence the outcome of risk controversies. While all European countries seem to follow a consensus-orientated approach, some are more sensitive to environmental issues than others. It has been argued that the representation of environmental interests at the political level makes a big difference (Dryzek 1997; Vogel 1993). Taking energy efficiency and per capita emissions as examples, the top performing countries (Germany, Japan, the Netherlands, Norway and Sweden) are all corporatist countries with an institutional representation of environmental interests.

Returning to the question of how we select risks, it seems useful to look at the contribution of cultural theory (Douglas and Wildavsky 1982). Cultural theory has argued that there exists a link between cultural bias and action, especially between one's perception of nature and decisions related to it. In a study on ecosystems managers, Holling found that when confronted by the need to make decisions with imperfect knowledge, they assumed that nature behaves in certain ways (cit. in Adams 1995:33). He reduced these assumptions to three views about nature. In the first, nature is perceived as benign, in the second as ephemeral and in the third as perverse/tolerant. Douglas (1988) and Thompson *et al.* (1990) added a fourth – nature as capricious – to produce the well-known four-fold typology.

Cultural theory further postulates that these four myths about nature corres-pond to four different social positions or types of persons: the individualist, the egalitarian, the hierarchist, and the fatalist. Individualists are enterprising people, relatively free from control by others, who strive to exert control over their environment and the people in it. Nature is seen as the benign context of human activity, not something that needs to be managed. Egalitarians have strong group loyalties but little respect for externally imposed rules, other than nature's. Here nature is fragile, precarious and unforgiving. Hierarchists inhabit a world with strong group boundaries and binding prescriptions. This view of nature combines the first two: within limits, nature can be relied upon to behave predictably. It can and should be managed, but one has to be careful not to push too far. Finally, fatalists have very little control over their own lives. They are assigned to their fate and see no point in changing it. For them nature is unpredictable. There is no point in intervening (adapted from Adams 1995).

The strength of cultural theory is that it points to the importance of cultural bias which helps to select courses of action and reduce the complexity of the issues involved. Views about nature seem to be particularly influential in guiding actors to define risk problems. However, the contribution of cultural theory has its limits. It has been noted that the model is too general and therefore does not lend itself easily to analyse specific problems. For example, it is not clear how people acquire their world-views: are they innate? Can they change? Neo-institutionalist approaches might be better adapted to provide a more fine-grained analysis (O'Riordan and Jordan 1999). Furthermore, the four-fold typology seems unduly complex since both hierarchists and egalitarians want state control, while indi-vidualists want less. It thus seems to me that we should adopt a more parsimoni-ous model, in which two opposing types of actors based on different attitudes towards risk are the main protagonists and a group of (undecided) bystanders observes the disputes, eventually taking sides. This approach remains agnostic about the ways in which world-views relate to positions in social structure. It stresses the important point that at the beginning of risk controversies only a minority takes a position.

So, should we trust the experts? Even before Western societies witnessed the emergence of risk controversies on a large scale, Weinberg has pointed out that 'there are answers to questions which can be asked of science and yet which cannot be answered by science' (Weinberg 1972: 209). Funtowicz and Ravetz (1992) have argued that formal models (risk statistics) are inadequate since they do not help to deal with uncertainty. Scientists as lay persons are operating within a value-laden context where facts are uncertain, values in dispute, stakes are high, and decisions urgent. Experts are amateurs when uncertainties exist. Funtowicz and Ravetz argue that we therefore need a dialogue among all the stakeholders in a problem, regardless of their formal qualifications or affiliations, an 'extended peer community'. It is because the risk calculus cannot be applied that experts do not have a better grasp than lay people. Lay people should therefore have the same say in these decisions as experts.

It seems evident that natural scientists do not like these propositions; yet, furthermore, they do not like the underlying problems either, since they pose too many demands on them. First, scientists are generally looking for simple yes–no answers to well-defined questions and try to avoid ambiguous results. Second, their expert knowledge is mainly limited to a small specialised field that does not allow them to include other approaches that could contribute to the solution of a problem. In judging risks, scientists should have a competence of judgement that goes beyond the yes–no routine of their daily lab practice but at the same time transcends their narrow specialities (Marcus 1988). How scientists reacted to these challenges will be discussed in Chapter 3 in greater detail.

Politicians may not like the thought that they have to decide issues which are unsolved scientifically, and yet have been put on the political agenda. Politicians, like experts, therefore try to 'educate' the public by pointing out how low statistical risks 'really' are. More recently, they have even taken on the message of risk sociology and stress that we cannot get 100 per cent safety. For a long time, the myth had been established that the natural sciences have in principle to offer more robust knowledge than the soft social sciences. However, in risk controversies, as in new research frontiers more generally, there is no uniform scientific answer or body of knowledge. Instead, there are different responses in different research cultures and national jurisdictions. This comparative dimension will be developed in Chapters 4 and 5.

A network approach

How can the aspects of policy making be brought together with scientific disputes? How can we envisage international cooperation, which to a large degree depends upon different national interests and the state of knowledge, which may be different in different countries? As noted earlier, mono-disciplinary or reductionist approaches are insufficient to come to grips with a multi-faceted object like this one. I have adopted a network model that combines approaches from several fields of research: the literature on policy networks, the literature on discourse coalitions, the theory of actor networks and, more generally, from literature in the field of the sociology of scientific knowledge (SSK) and international relations (IR), especially its constructivist strand.

The policy networks literature has stressed the idea that informal relations between key actors are necessary to initiate and implement specific political aims (Heclo 1978). Discourse coalitions can be conceived of as policy networks around issues in which the discursive, public debate occupies an important role (Dryzek 1997; Hajer 1993, 1995). Actor Network Theory has convincingly argued that non-humans (nature, technology) make a difference (Callon 1987; Latour 1990). It has subsumed all kinds of agency under the concept of 'actant'. However, by so doing, it loses analytical precision which can be achieved by distinguishing between actors, strategies and resources.

The aim of this case study is not to measure the explanatory power of single variables. Rather, I shall demonstrate that there is a plausibility of a specific explanation that has wider relevance. The argument is that an institutional approach helps to best understand the process. It is evident that with regard to global environmental problems several traditional modes of explanation offer too little for an adequate understanding. *Pluralist* approaches focus on the interplay between constitutional bodies (like parliament, government and administration). *Neo-corporatist* approaches emphasise the role of powerful interest groups and their influence on policy making. *System theoretical* approaches postulate the fragmenta-tion of social spheres and their closure. And *rational choice* theory starts from the assumption of rational individuals calculating costs and benefits. In IR literature, this amounts to identifying positions of 'national interest' that can be objectively defined. In contrast to these approaches, *institutional analyses* focus on decision rules, path dependency and structuration processes (Arthur 1990; Giddens 1984; O'Riordan and Jordan 1999). Institutions are the product of human choice, yet constrain and shape behaviour. Rules of the game and rules of appropriateness encapsulate values, norms and world-views, thus guiding behaviour and supplying a repertoire of legitimate aims.

Policy networks

Networks are usually defined as an institutional form that transcends market and hierarchy as 'principal' governance structures. They are constituted by actors from many different backgrounds and levels of hierarchy. Policy networks in the field of environmental policy include scientists, representatives from government and administration, non-governmental organisations (NGOs), international organisations and industry. In addition to the more general proposition that policy networks can solve collective action problems in cases where markets and hierarchies are absent or fail, I address the special problem of decision processes under conditions of uncertainty. In negotiation systems, this means that actors do not have the information necessary to calculate their own pay-off, much less the pay-offs of other actors. In situations of uncertainty, we orientate our behaviour on the reputation of people we trust. We trust them either because we have dealt with them in the past or because they show 'seals of approval' (Klein 1997).

The policy network approach adopted here looks at informal relations between actors. Generally speaking, networks can mobilise actors and resources. The latter can be material (money, technology, laboratories) or symbolic (scientific know-ledge, rhetoric). Ideas provide rallying points around which allies cluster. This pre-structures the conflict and maps potential allies onto the political stage which will be enacted through the unfolding of the real process.

Figure 1.1 shows the essential elements of my approach in schematic form. The left part denotes the structural properties of the policy domain, decisions under uncertainty and the problem of representing diffuse interests. Speakers may com-

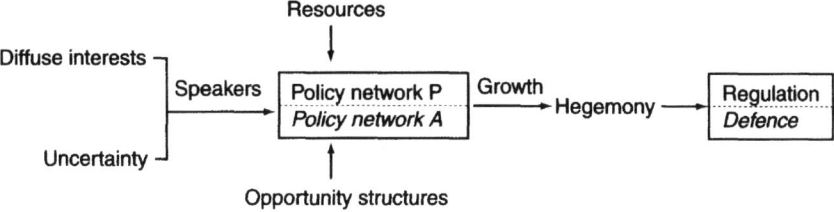

Figure 1.1 Policy networks approach

bine both aspects and speak up as advocates of diffuse interests. In all likelihood, they are opposed by speakers of well-organised interests. These speakers represent policy networks, the former the pro-regulations network P, the latter the anti-regulations network A. Conflicts over health and environmental issues usually arise over problem definitions and alleged causal relations. In this, information, scientific data and their interpretation are primary resources (Beck 1996; Hannigan 1995). From this results a struggle for scarce (symbolic) resources. Their availability enhances or constrains a policy network's opportunities for growth. Public credibility is one of the most valuable resources. Institutional opportunity structures (political constitution, policy style, institutional environment, institutional decision rules) may also enhance or constrain the growth of a policy network. If one policy network gains hegemony, stable political regulations are likely to follow. In what follows, I address each step in turn.

Policy networks and their speakers

Kenis and Schneider (1991) stress the usefulness of the network concept mainly for those policy processes where not only formal institutional arrangements are important but above all informal, often decentralised decisions in which complex actor constellations and resource dependencies play a role and where actors rely on reputation and trust:

> A policy network is described by its actors, their linkages and its boundaries. It includes a relatively stable set of mainly public and private corporate actors. The linkages between the actors serve as channels of communication and for the exchange of information, expertise, trust and other policy resources. The boundary of a given policy network is not in the first place determined by formal institutions but results from a process of mutual recognition of functional relevance and structural embeddedness.
>
> (Kenis and Schneider 1991: 41f.)[10]

Powell has emphasised the aspect of trust in the following way:

> [T]he most useful information is rarely that which flows down the formal chain of command in an organisation, or that which can be inferred from

price signals. Rather, it is that which is obtained from someone you have dealt with in the past and found to be reliable.

(Powell 1990: 304)

Structures of interaction remain often hidden to the outside observer. There is no formal membership in such networks but there are common policy goals that are partly defined by ideological elements. Hugh Heclo coined the notion of the *issue network* for such networks in which there is a fluctuation of members; no one has complete control over the network and material interests are subordinated to intellectual and emotional engagement. 'Network members reinforce each other's sense of issues as their interests rather than (as standard political or economic models would have it) interests defining positions on issues' (Heclo 1978: 102).

Paul Sabatier has emphasised the antagonistic dimension of policy processes. He conceptualises policy change in a model of interaction of rival advocacy coalitions within a policy field. Allies are actors of public and private institutions that belong to different levels of hierarchy, share common values and fight for their realisation. In order to do so, they try to manipulate the rules, the budget and the employees of government administration. Policy change depends also upon environmental factors outside the policy field, and upon stable system parameters within the policy field that provide opportunities and constraints (Sabatier and Jenkins-Smith 1993: 5). Applied to risk controversies, this means that regulations are unlikely in the absence of skilful advocates which mobilise diffuse interests, above all by alarming the public. In so doing, they attack the opponents of regulation, accuse them publicly of neglecting highest values like health and safety, and put them on the defensive (O'Riordan 1971). As soon as diffuse interests are represented by advocates, the well-organised interests, too, have to send speakers into the public arena. Thereby the structural advantage of well-organised interests is relativised. In public debate, arguments are subject to the need to present themselves in such a way as to be generally acceptable. Majone (1996b: 610) expresses this mechanism in the following way:

Because policy is made out of language, arguments are used at every stage of the process. Every politician understands that ideas and arguments are needed not only to clarify his position with respect to an issue, but to bring other people around to his position. Even when a policy is best explained by the actions of groups seeking selfish goals, those who seek to justify the policy must appeal to the intellectual merits of the case.

Whenever speakers take sides for the two alliances in public, the questions 'How many reputable scientists support a position?' and 'How well founded is it?' arise. If a large number of scientists supports a specific position, the resulting effect can greatly increase credibility and even be taken as an indicator of consensus. In the current discussion on climatic change, the fact that more than 2,000 atmospheric scientists worldwide see a discernible human-made change in the climate was a major factor lending credibility to their claims. Likewise, the reputation of a research institution will influence the credibility of its claims. However, expert opinions will always be challenged. As judgements offered by experts claim to be based on scientific data and theories, part of the dispute is about what constitutes 'good' science – or what can be considered to be science in the first place.

Social dynamics

In self-reinforcing social processes (Maruyama 1963) there is a circular stimulation that affects the actors' motivation. Such processes create effects that they use again for their perpetuation. They also tend to involve the environment in their own processing: 'For whatever reasons other actors from outside may get involved into self-reinforcing processes – they run the risk to be drawn into the fray themselves' (Mayntz and Nedelmann 1987: 665). The circular stimulation of the motivation of social actors is a distinctive feature of the case under discussion.

Industry finds itself in a dilemma since discourses unfold in a potentially autonomous and dynamic way. On the one hand, it fears these dynamics because the consequences are hardly foreseeable. On the other hand, it sees the need to engage in the public discourse as early as possible in order to define the problem in such a way as to favour its interests (Lau 1989a: 389f.; cf. also Gehring 1994: 219). Scientists who are critical towards industry usually have more credit in public since they seem to be more honest and trustworthy than their counterparts who work for and are paid by industry. The latter are sometimes perceived as hired guns who deliver a partisan opinion disguised as science – for cash (Harvey 1998; Press and Washburn 2000). But no matter if friend or foe of industry: each side casts doubt on the quality of the other side's claims. Latour has made a comparison between scientific controversies and the arms race. This is quite illuminating as the feedback loop is quite similar. As soon as one side has established a knowledge claim as hard fact, the other side has to draw equal – or else submit: 'The costs of disagreeing will increase.'

> Positive feedback will get under way as soon as one is able to muster a large number of mobile, readable, visible resources at one spot to support a point . . . Once one competitor starts building up harder facts, others have to do the same or else submit.
>
> (Latour 1990: 34f.)

Networks in the sense used here are similar to Sabatier's advocacy coalitions and Haas's epistemic communities (Haas 1992; Sabatier 1993). Their members have similar world-views, assumptions about cause and effect relations and proposed solutions. In other words, they are networks of like-minded people. Unlike network models in which heterogeneous actors reach agreement within a network through the exchange of resources (Kenis and Schneider 1991; Mayntz 1993), this route is obstructed by the ideological orientation of the actors.[11] Here, two rival networks are battling for hegemony. Particular attention is given to the following mobilisation processes and their linkages:

- the mobilisation of non-material resources;
- mutual reinforcement of allies and resources;
- linear and non-linear mobilisation of undecided actors;
- growth of one network at the cost of another.

A network may mobilise resources and win new allies independently of one another, as well as gaining new resources through new allies and allies through new resources. (In what follows I shall be concentrating on positive feedback processes. The exposition also applies, with the sign reversed, for negative feedback processes, cf. Maruyama 1963.) The growth of a network is linear if the same number of actors or environmental resources is mobilised during comparable intervals. Growth is non-linear if a number of actors (or resources) suddenly join the network. In antagonistically structured areas of policy, where two camps are facing each other in a battle for hegemony, there is a further possibility of non-linear growth through attacks by one network on its opponents, thereby winning over allies and/or resources from the opposing network. This mechanism (growth of one network at the cost of another) results in lasting and decisive changes to the balance of power between the networks. In areas of policy characterised by the existence of two opposing coalitions, success by one side means the collapse of the other. If one side succeeds in releasing allies or resources from the opposing alliance and recruiting them to its own cause, the momentum of this process may bring about a domino effect or even a chain reaction, especially where a large population of actors outside the two networks has remained undecided for a long time but is suddenly mobilised. This is particularly likely when opinion leaders – actors by whom large numbers of other actors are influenced – cross over from one camp to the other (Katz and Lazarsfeld 1955).

The probability that a position of hegemony will be achieved is increased if powerful actors and resources are recruited to one of the two sides. Powerful actors can be defined by reference to the resources at their disposal. If material resources alone are the decisive factor, this definition is unproblematic and industry can be regarded as a powerful player. Material resources are a major element in the production of symbolic resources, and are thus a necessary condition for both sides in contesting the controversy, for example, scientific laboratories and infrastructure. Dedicated scientists, public interest groups and policy entrepreneurs can take on the role of speakers and play a decisive role in mobilising public opinion. If a government or business is faced with a dramatic loss of credibility, it is likely that it will withdraw from the argument or even switch sides. Where symbolic resources are important, industry cannot be ranked a priori as a powerful actor, although here too it can be assumed that it will be mainly businesses that have advantages over other actors in terms of resources and information. As public pressure might influence the shaping of policies and can also affect business policy through changes in consumer preferences (including boycotts), a swing in public opinion may diminish industry's powerful position. Actors in a position to mobilise public opinion by means of symbolic resources deserve special attention.

Reputation

Where decisions are taken under uncertainty, it is principally symbolic resources that make a decisive difference: scientific scenarios and warnings, interpretations of the situation and proposed solutions. 'Information' is not a good term to describe these. Knowledge is not just information but interpretation, judgement and understanding. Institutions that are recognised for their manner of interpretation and judgement produce seals of approval (Klein 1997). It is characteristic of this process that many of the actors (and new allies) are corporate actors rather than individuals, representing an institution and acting on its behalf, without taking orders from them every step of the way or being under their control (Coleman 1982; Mayntz and Scharpf 1995: 50f.). A diplomat who took a decisive role in the international agreements on the ozone layer expresses the concept as follows (Lang 1994: 174; see also Putnam 1988):

> The individual negotiator in many instances may be much more than a 'puppet on a string': he or she may enjoy a certain leeway within the instructions they have received; he or she is supposed to use their personal skills to persuade the other side, extract concessions from the other side etc. Thus the personal factor, the professional and cultural background of a negotiator have their impact on the course of negotiations.

In the literature, these representatives of companies and public authorities are conceptualised as corporate actors within the principal–agent theorem (Coleman 1985; Pratt and Zeckhauser 1985). They represent an institution and act as a deputy for them without being controlled for every single step they make. This logic also applies to scientists and journalists. This 'fuzzy relation' makes it possible that – to give an example – an actor's statement gains enhanced visibility and reputation if a prestigious institution is implied. If a scientist representing a major university or a prestigious institution takes a position in a controversy, this is attributed not only to the speaker personally but also to the institution that employs him, which means that such public statements get a boost in symbolic significance. If a Nobel Prize winner or a NASA scientist publishes data that contradict data deriving from a scientist at an unknown university, credibility is likely to be skewed in favour of the former. Scientists of highly renowned institutions have better chances of being heard, and above all, believed than scientists of small provincial universities or labs (cf. Crane 1965; Merton 1973c).[12] A similar mechanism is at work within the media. If a journalist takes sides in such a controversy, this is taken to be a position of the paper at the same time. The more prudent and objective the paper appears in general, the more trustworthy is such a partisanship.

Framing: the role of ideas

Environmental hazards that are abstract (those which cannot be seen, felt, smelled or tasted) call for symbolic representation. In this, ideas can pre-structure a policy option, serving as symbolic resources to lend it credibility and thereby increase its attractiveness (Kingdon 1984: 131–4). In disputes about safety, health and the environment two principal positions based on two different norms can be identified: a risk taking and a risk averse position. The first is based on the norm that the public has to be protected against dangers; the second that no interference with private initiative should be allowed. The first position advocates quick regulations, even if there is a lack of scientific understanding. The second wants a comprehensive review process and the proof of a causal mechanism before regulations are seen as legitimate. Both positions are based on principled, normative orientations that are not open to negotiation: the precautionary and the wait-and-see principle (cf. Milbrath 1984; O'Riordan and Wynne 1987; Otway and Thomas 1982). The appeal to norms can constitute a powerful instrument to persuade actors of a specific policy ('It may be easier to seduce a Communist or a Christian than to bribe him [*sic!*]', as Elster [1989: 13] remarked).

Recently, 'constructivist' political scientists have discovered the role that ideas, interpretations and scientific expertise play in the policy process (Fischer and Forester, 1993; Goldstein and Keohane, 1993; E. Haas 1990; P. Haas 1992a, 1993; Majone 1989, 1996b; Sabatier 1993).[13] However, the mainstream of 'realist' political science is generally sceptical about the role of ideas in the political process. There seem to be two contradicting reasons: one questions the power of ideas, the other acknowledges their power but sees it as detrimental. Interest-based paradigms explain policy processes and results by the self-interest of the actors. The alleged influence of intellectual or scientific knowledge within the policy game is thus quickly dismissed as a 'naively idealistic' view. The alleged influence of science is especially seen as problematic since it refers to a techno-cratic policy advice model, which has shown its limits time and again. As several authors have argued persuasively, scientific data are normally instrumentalised for pre-existing political options (cf. the garbage-can model, Cohen *et al.* 1972; Kingdon 1984; Shils 1987). The same logic seems to be at work in political decisions under uncertainty when two competing definitions of the situation and two corresponding political proposals exist; in such controversies interest groups cluster around scientific hypotheses, scenarios and explanations. 'Whatever political values motivate controversy, the debates usually focus on technical questions' (Nelkin 1979: 16; cf. Collingridge and Reeve 1986). Little wonder that industries under pressure are nearly always on the side of those who deny or downplay the risks. However, this position is often supported by actors that do not have an immediate stake in the issue in that they do not themselves profit from the maintenance or introduction of a possibly risky technology. Likewise, opponents of such a technology may define their position not by recourse to self-interested motives since the potential damage is distant in time or space. In such controversies norms and ideas exert a decisive motivating power. The result is that we not only get conflicts of interests but conflicts of values, too.

Firms will usually take the wait-and-see position and one could think that they hide their special interests behind commonly accepted norms. Although this is true much of the time, there are two reasons to treat norms as real and autonomous.[14] The first is that apart from industry, other actors defend private initiative, entrepreneurship and the wait-and-see principle, i.e. demanding scientific proof before action becomes seen as legitimate. The second is that industry is binding itself through public engagement and commitment. It is appealing for industry to hide behind generally accepted norms (which go beyond their specific narrow self-interest when demanding scientific evidence). Public support may be enhanced by so doing. The rub, of course, is that this strategy may turn out to be self-defeating in the long run.

Ideas can be analytically distinguished as shown in Figure 1.2. They serve as symbolic resources or are part of world-views, and thereby pre-structure a discourse. Take symbolic resources first. Here, one can further distinguish two forms of symbolic resources: scenarios and alarm signals. Scenarios are forecasts of future events which are calculated on the basis of more or less known input data.

Scenarios may be seen as virtual alarm signals. Real alarm signals, however, are much more effective. Heiner (1986) sees the *signal-to-noise ratio* as an important variable for the probability of preference change of actors. Both are used in the political process as a resource and are apt to resolve deadlocks. Real disaster warnings or crises trigger swift changes by causing relevant actors to reach a common interpretation of the situation and initiate appropriate measures quickly. Alarm signals do not develop their effect of themselves, but must first be issued by speakers. That is, speakers are engaged in the collection and interpretation of data, which are then introduced into the policy process in the form of symbolic resources. From world-views norms are derived about the 'right' relation to nature (and uncertainty), and definitions of the situation or solutions to problems are developed (Gehring 1994; Goldstein and Keohane 1993; Haas 1992a, 1993;

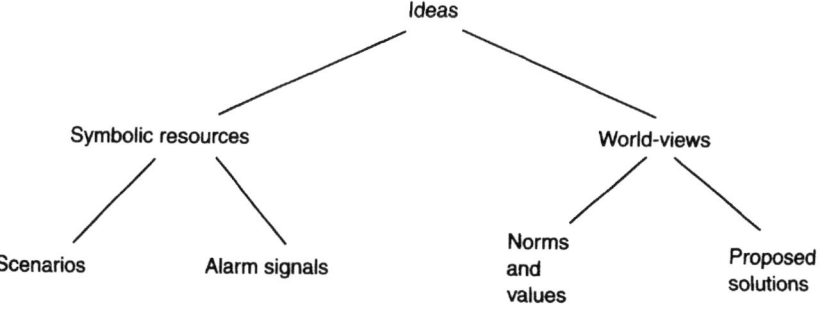

Figure 1.2 Typology of ideas

Lau *et al.* 1991). They help to

1   outline issues and problems not previously on the agenda,
2   make a selection from a multitude and complexity of options and
3   specify ways to a solution.

If ideas become so important, the appropriate research strategy is to analyse these policy arenas (Renn 1992) and the claims making activities in them. We identify key actors and their claims, analyse how they frame problems, how they put them into media packages and develop story lines (Gamson and Modigliani 1989; Goffman 1975; Hajer 1995; Rein and Schön 1993). Whereas ideas are treated with 'realistic' approaches based on power or self-interest as epiphenomena, their role here is rated more highly, though not so highly that they are developed into a monocausal explanation. However, they are instrumental in determining actors' preferences. When decisions are made under uncertainty, we expect that the precautionary and wait-and-see principles oppose each other. The better one of these groups succeeds in framing the situation, the more likely it is that their position will appear as more trustworthy within the (small) actor population and the broader public (Jones 1994).

However, as ideas gain in importance, cooperation between contending camps is less likely to emerge. This is the basis for the second type of scepticism about the influence of ideas on the political process. If 'passions' and ideologies play an important role (Hirschmann 1977; Mayntz and Scharpf 1995: 54ff.; Pizzorno 1986), the rational interest mediation through negotiation is thwarted. [15] Ideas are thus seen as part of the problem, not part of the solution. However, in negotiation systems where problem-solving and not interest mediation is the main goal, ideas can facilitate cooperation (Benz 1994; March and Simon 1958: 150f.; Mayntz 1993: 47f.).

In Table 1.2, three different modes of conflict resolution are shown. The

Table 1.2 Modes of conflict resolution

		Conflict between special interests	
		STRONG	WEAK
Conflict between cognitive orientations	STRONG	Deadlock	Integrative bargaining/ **Exceptions, concessions**
	WEAK	Bargaining/ **Compensation, arithmetic**	Technical Problem-solving/**Adaptation clauses/standards**

instruments available to each mode are printed in bold typeface. I start with the lower row, which might be called the 'standard view' since it excludes strong cognitive orientations. In the standard view, we have bargaining and technical problem-solving. Compensation payments and standard setting are typical instruments. If we follow the constructivist strand in political science and include the dimension of cognitive orientation (upper row), we can assign deadlock a special place (the north-western quadrant) and identify an additional mode of conflict resolution: integrative bargaining (IB), denoting the combination of bargaining and problem-solving (cf. Pruitt and Lewis 1977; Walton and McKersie 1965; Young 1994).[16]

To escape deadlock, three possible paths seem possible: first, a bargaining process, second, technical problem-solving and third, integrated bargaining. To go from deadlock to bargaining, strong cognitive orientations have to be overcome. To go from deadlock to technical problem-solving, actors have to give up (or tone down) simultaneously their special interests and cognitive orientations. However, during technical problem-solving it is unlikely, indeed 'almost unthinkable', that 'delegates proposing a specific standard would offer side payments to opponents in order to compensate for the economic disadvantage they could incur from the standard,' as Schmidt and Werle (1998: 283) put it. Negotiations are perceived as being conducted behind a 'veil of ignorance', thus leading to just outcomes (Rawls 1971).

Finally, to go from deadlock to integrative bargaining, actors switch from pursuing their partial interests to a common good orientation. I argue that this is likely if the common good orientation becomes dominant. IB has the advantage of allowing a switch to both bargaining and technical problem-solving, because it allows for a much wider range of exceptions than the other modes. Bargaining is facilitated by side-payments, technical problem-solving by means of adaptive clauses (cf. Scharpf and Mohr 1994). IB can combine side-payments with adaptive clauses. IB thus makes the 'almost unthinkable' possible, i.e. that parties in favour of a specific standard or regulation would offer side-payments to opponents in order to compensate for the economic disadvantage which would result from the standard. The experts devising the standard may not be identical to the delegates participating in the bargaining process. However, since both adhere to the same policy network, the relevant information will be available to key actors.

Timing

The two camps battle for hegemony in a long-lived controversy. Over time there will be important shifts in the balance of power between the camps, and the structural dimension must therefore be considered along with a *dynamic* one. The time dimension is a critical element in the analysis. This means that the chronological order of events constitutes an 'order of things', an irreversible new reality. After a window of opportunity has closed, an option may be unavailable for a very long time to come. Actor strategies have thus to reckon with issues of timing and synchronisation (Adam 2000). Methodologically, this means that the sequence of events is of crucial importance for the explanation of the outcome of the controversy. This path-dependency is an important feature of environmental issues and will be thematised throughout the study.

Social constructivism

The issue of claims making and the reference to the literature about the sociology of scientific knowledge (SSK) raise troublesome questions of epistemology and political affinities. Here, as elsewhere, relativism is thought to show its nasty face (Burningham and Cooper 1999). In particular, it has been argued that social constructivism is bound to downplay environmental problems. Canadian sociologist Raymond Murphy has advanced the argument that social constructivism, as practised in SSK, is pre-ecological and does not acknowledge the embeddedness of social action in nature. Furthermore, he assumes that only a realist ontology can protect us from being recruited by the 'risk deniers' (cf. Beckerman 1995; Simon and Kahn 1984). Murphy perceives sociologists mainly as being interested in debunking reports about environmental problems as 'social scare stories'. He explicitly states: 'If social constructivism did not already exist, it would have been invented in order to oppose the changes in life style and in profitability needed to reduce the degradation of the natural environment' (Murphy 1997: 56).

This is at odds with constructivist positions as developed by MacNaghten and Urry (1998), Yearley (1996), Hannigan (1995) or Wynne (1996a). In their view, knowledge about nature is socially constructed and culturally contingent. Here, authoritative scientific knowledge is a product of negotiation between scientists and policy makers, as they have exemplified in case studies about BSE or climate change. They point out that sociologists should challenge the technical and natural sciences by disclosing the social and cultural assumptions upon which they rely.

Wynne (1996a: 363) agrees that 'sociological deconstruction of knowledge may find itself in unwelcome company, politically speaking'. Moreover, he acknowledges that SSK has generally been simplistic in conceptualising political processes: 'It has adopted a model of public issues as constituted by interaction and competition only between actively chosen stances that reflect real interests; thus the issues themselves are not problematized beyond identifying the "hidden" interests of the protagonists' (Wynne 1996a: 366). Taking the example of climate change science, which he describes at length, Wynne admits that

> these sociological observations about the scientific knowledge of global warming could of course contribute to a deconstruction of the intellectual case for the environmental threat, and thus also to a political demolition of the 'environmentalist' case for internationally effective greenhouse gas controls.
>
> (Wynne 1996a: 372)

Here he mentions the fears of Dunlap who – like Murphy – states that to show how science constructed global environmental problems amounts to suggesting that they do not really exist. But Wynne disagrees: 'The point is *not* thereby to throw out the knowledge; it is to debate the social assumptions, which first need to be identified.' Wynne also shows that attempts at deconstructing knowledge claims came from climate scientists themselves, without encouragement from SSK. This suggests that scientists themselves are aware, at least in crucial moments of controversies, that what counts as a 'fact' is not found, but made.

Hannigan (1995) makes it most clear that a constructivist position need in no way undermine a concern for the environment. He follows up a debate in the 'social problems' literature and refers to a paper by Spector and Kitsuse (1973). They argue convincingly that social problems (like drink driving, child abuse) are not objectively 'out there' in reality. Rather than triggering attention by being 'objective facts', they are the product of claims making activities. Hence the roles of claims makers, scientists, the media, environmental pressure groups and politicians need to be analysed. Hannigan draws an analogy to environmental problems and argues that, rather than being observable in reality, environmental problems have to be established via claims making activities.[17] Rather than undermining a concern for the environment, it has to be noted first of all that all claims are embedded in social action. From this, constructivists draw the following conclusion: '[D]emonstrating that a problem has been socially constructed is not to undermine or debunk it, since both valid and invalid social problem claims have to be constructed' (Yearley cit. in Hannigan 1995: 30). The chances for reception of such claims depend upon a variety of conditions which Hannigan (1995: 55) lists:

1   Availability of scientific data and scientific authority;
2   Existence of popularisers or advocates who bridge environmentalism and science;
3   Media attention: the problem is seen as novel and important;
4   Dramatisation by symbolic and visual means;
5   Economic incentives;
6   Institutional sponsor who lends legitimacy.

As Ungar remarked,

> Recognition in public arenas, which is the sine qua non of successful social problems, cannot be reduced to claims-making activities, but depends on a conjunction of these and audience receptiveness. Claims-making, after all, can fall on deaf ears or meet bad timing.
>
> (Ungar 1992: 484)[18]

Data and interpretation

The contribution of SSK to the understanding of scientific controversies is that it conceives of scientific knowledge as being produced in a closure process that has two main inputs: *data* and their *interpretation*. Laboratory data are the product of a mobilisation of nature (Latour 1987: 94ff.); interpretation is the product of mobil-isation of bias (Schattschneider 1960). There is a close link between the two. The production of lab data starts from preconceptions (bias) and the mobilisation of bias often occurs with the help of empirical evidence (data). It is evident that bias which goes into scientific data is problematic. It has to disappear in the final scientific claim if it is to gain the status of certified knowledge.[19] Only nature is supposed to be the judge about competing knowledge claims. This indicates that we trust data more than interpretation. However, data can only be produced with the help of interpretation: every research starts with some implicit assumption, preconception or bias. The collective displacement of interpretation can be demystified precisely by analysing claims making activities in which allegedly no interpretation did enter at all. The natural sciences are especially apt for such an exercise.

Are we to conclude from all this that scientists engaged in such controversies, because of being biased one way or the other, are doing 'bad science'? Not at all, as sociologist of science H. M. Collins has argued. According to him, social mechanisms by which scientific controversies are closed do not make the results unscientific. He even says that:

> some 'non scientific' tactics *must* be employed because the resources of experiment alone are insufficient . . . Nevertheless, the outcome of these negotiations, that is, certified knowledge, is in every way 'proper scientific knowledge'. It is replicable knowledge. Once the controversy is concluded, this knowledge is seen to have been generated by a procedure which embodies all the methodological properties of science. To look for something better is to grasp a shadow. Scientists do not act dishonourably when they engage in the debates of typical core sets; there is nothing else for them to do if a debate is ever to be settled and if new knowledge is ever to emerge from the dispute. There is no realm of ideal scientific behaviour. Such a realm – the canonical model of science – exists only in our imaginations.
>
> (Collins 1985: 143)

The objectivity of scientific results is obtained mainly through an annihilation of the motives, strategies and interests that may have played a role in the produc-tion of knowledge. The influence of scientists is explained by the thus achieved purity of results. Context and content of science are created in two different but complementary processes that are simultaneously taking place. Latour calls them hybridisation and purification. In a similar way, Luhmann (1989) talks about extra-scientific influences which are treated like disturbing noise and thus elimin-ated. Nevertheless, this noise – which includes personal idiosyncrasies – is an indispensable requirement for the functioning of research. Likewise, Jasanoff (1986: 229) observed that:

> the experts themselves seem at times painfully aware that what they are doing is not 'science' in any ordinary sense, but a hybrid activity that combines

elements of scientific evidence and reasoning with large doses of social and political judgment.

The hermeneutic triangle

Compare this with the widely shared position that holds that there is a principal difference between the humanities and the social sciences on the one hand, and the sciences on the other (e.g. Gadamer 1960). In the former, all important discoveries have been made a long time ago. There is nothing new under the sun. Research in this field can only be about a different interpretation of the same statements and things. The knowing subject and the object of inquiry are caught in a 'hermeneutic circle'. The natural sciences are another matter: through their observations of the world an increase in knowledge is possible. The 'new' sociology of science has cast doubt on this picture. The point is often made that every field of knowledge is made up by data and interpretative problems (Holton 1994). Furthermore, facts do not exist independently of a community of scholars who subscribe to specific sets of assumptions. Knowledge, or 'facts', are more accurately understood as consensually accepted beliefs than as proof or demonstration (Fischer 1998). To illustrate this, I shall replace Gadamer's metaphor of the hermeneutic circle with the hermeneutic triangle. This is defined by three points (A, B and C in Figure 1.3): A is a researcher convinced of her own results. She stresses the objectivity of her method and the evidence provided by the data. B is the charitable critic who trusts the data basically but sees a somewhat different relation. He probably rates the data somewhat lower than A but higher than the malicious critic C. Likewise, B thinks that there is more interpretation than A thinks, but less than C thinks. On the basis of A's data, B may even support A in a way that was not seen by A (see position B' in Figure 1.3). C is the malicious critic. He opposes A's position by taking one of three possible strategies:

1   He disputes the validity of the data.
2   He disputes the relation between data and interpretation.
3   He disputes the interpretation.

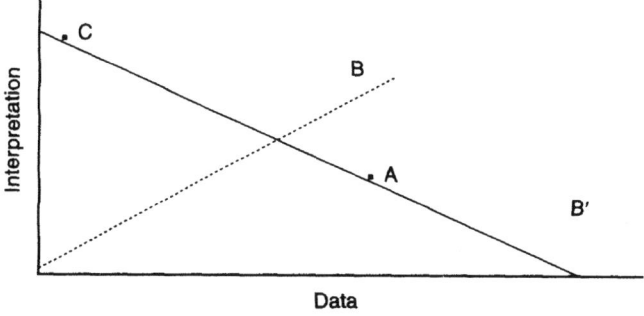

Figure 1.3 The hermeneutic triangle: distribution of knowledge claims supports A.

In all three cases the data do not prove what A claims. C thinks A's results are over-interpreted, maybe pure speculation. If we represent the three scientists graphically, we get the picture shown in Figure 1.3.

This graph illustrates the scientific community's judgement about A's knowledge claims: it depends on the quantitative distribution between the three positions and their relative proximity to A or C. If we assume that A is a single researcher, we can identify two possibilities:

1   If B dominates over C, A will receive a net support even if B does not share all of A's statements. This is likely in cases where B is a group that is composed of individuals $(B_1, B_2, \ldots B_n)$ of which a large fraction or several highly reputed scientists are near to A (including the possibility that position B′ provides a 'stronger' version than A).

2   If C dominates over B, A's knowledge claim is not accepted: it has to be regarded as 'refuted' (Figure 1.4).[20] This is likely in cases where B is a group that is composed of many individuals $(B_1, B_2, \ldots B_n)$ of which a large fraction is near C. In this case even the most charitable critic is closer to C than to A.

Scientific facts only emerge if scientific results are found to be trustworthy by relevant peers, even if they do not repeat the steps of this process of knowledge production. As an extending chain of convinced adepts certifies the accomplishments (of an experiment, for example), a fact becomes established. Shapin (1984) coined the term *virtual witnessing* for this process. In risk controversies the involved parties will argue about the trustworthiness of data and their interpretation. The hermeneutic triangle can thus be applied to the broader social and political debate as well.

Institutional opportunity structures

Policy networks do not develop in a vacuum. They are embedded within an institutional structure of a country, a region or the international system. Their chances to succeed are contingent upon institutional opportunity structures and

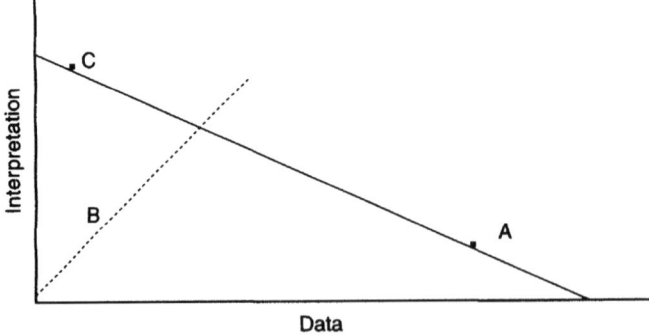

Figure 1.4 The hermeneutic triangle: distribution of knowledge claims undermines A.

resources that may both be influenced by historical factors. Such institutional factors are found in the political constitution, the policy style, and institutional decision rules. Decision rules allow the mobilisation of institutional bias, which is one of the most powerful resources in the policy game, as Schattschneider (1960) has emphasised. The definition of alternatives is seen as the highest resource of power. Imagine two decision rules for the regulation of toxic substances, one requiring scientific proof that substance X is harmful to human health and the environment, the other requiring a lesser degree of evidence to justify pro-active intervention. Uncertainty will be used to legitimise inaction in the first instance but it will be used to legitimise government control in the second. The existence of such rules does not determine how the issue will be decided but it provides advocates of either side with more or less powerful weapons. It also may be that the regulatory framework has different, even contradictory rules. Again, *framing* will make the difference. Lau *et al.* (1991) see the use of interpretation as the most common practice of manipulation in the policy process. They define interpretation as a bunch of propositions that have decisive influence over particular political decisions. 'The aim of each interpretation is to emphasise a dimension of judgement that will lead people to prefer one policy proposal over competing, alternative proposals' (Lau *et al.* 1991: 645). This is a good heuristic device for cases in which cognitive uncertainty reigns. It reintroduces the notion of power in a way which comes close to Lukes's 'third dimension' of power (Lukes 1974).

Historical contingencies lead to a path dependent development that is enabled or constrained by institutional structures. The existence of a network able to exploit such opportunities is decisive.

Institutional frameworks create different opportunity structures for the emergence of policy networks and their success. How does this relate to international cooperation? International control measures are influenced by the options of different national governments. International negotiations are shaped by the domestic policy game of the participating countries. Different political systems provide different institutional opportunity structures for diffuse interests. Several authors have argued that while well-organised interests are represented in corporatist countries, diffuse interests are unlikely to get onto the political agenda (Berger 1981; Martell 1994; Streeck and Schmitter 1985). This situation changes as soon as 'green' parties are in parliament (Vogel 1993) or green pressure groups mobilise public opinion. Generally, it may be said that corporatist regimes are better when implementing policies, since the close consultation between societal groups provides information, creates trust, and reduces transaction costs (Lotspeich 1998). This is to say that these countries do remarkably well once diffuse interests have been successfully represented or institutionalised (Dryzek 1997).

Regulation: international cooperation?

In risk controversies, more precisely: in decisions under uncertainity, decision-makers try to base decisions on the balance between present and future costs and benefits. Both are dependent upon the size of the damage that would arise if no action is taken now and the costs that regulation would entail. Both are uncertain and can only be estimated. This dilemma takes on different forms, depending on whether a government defends the interests of industry or those of consumers and the environment. The issue of international cooperation is tackled differently depending on whether a government protects its 'accused' industry or whether it aims to get votes from 'green' voters (Schneider 1988; Tsebelis 1990). I do not think a purely 'instrumentalist' view that sees the national government as executing the will of its industry is convincing (e.g. Levy 1996). In Figure 1.5 both courses of action are depicted graphically. If a government is under pressure from a risk averse public it will follow the upper path; if it tries to protect its industry, it will follow the lower path.

Scientific results, together with decision rules, are used to advocate or oppose regulations. They also help to mask other reasons (mainly economic ones). In the decision tree shown in Figure 1.5 it is clear that the potentially endless chain that would result when following the lower path is ideal to defend a reluctant industry since it amounts to a play on time. In case public pressure subsides, the topic may even disappear from the agenda without having bothered industry. Both the masking of vested interests and the play on time are appealing strategies for opponents of regulation. Still, regulation might come eventually, maybe even stricter than initially perceived.

Turning to the international dimension, it should be noted that, in contrast to the national level where a state authority exists to set the rules of the game and can enforce binding regulations, on the international level all actors have to agree upon common goals and on the means to achieve them. No one of them have enforceable means at their disposal; after all, a world government is neither possible nor desirable. The United Nations may play the role of a broker that establishes a framework of negotiations, mobilises technical and scientific expert knowledge and tries to coordinate the diverging interests. These attempts are more likely to be successful the more actors are involved in the process, develop a commitment and trust the broker. Through this self-binding mechanism the exit costs will rise and the actors try therefore to maintain their voice options (Hirschman 1970).

The argument of the global dimension does not guarantee that a global solution to the problem is sought after. Depending on the context, the argument may be used for or against international regulation. It can be used to mobilise public opinion in favour of regulations, especially if an industry of one country has had to accept regulations and consequently suffers a competitive disadvantage on the world market. It would therefore be entirely rational for them to export these

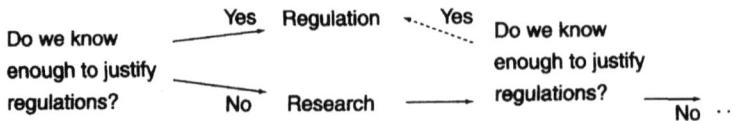

Figure 1.5 Political options.

regulations to other countries. The argument that the problem is global serves also to legitimise inaction, especially on the part of those countries that did not take action initially. This may lead to a situation where at the beginning of international negotiations only one side has a real interest in achieving binding and substantive measures. But even an industry that has been regulated in one country may speak out against international regulations if these would bring further tightening or if subsidiaries in other countries would be affected.

It is difficult to determine the preference of firms that face possible regulation a priori. For one, it is possible that industry favours regulations in other countries if this gives it a competitive advantage, in other words, if it can thereby increase its exports. But industry could rank solidarity higher and oppose regulations worldwide if they are perceived to pose a precedent that it feels has to be avoided in any case.[21] The chances of reaching international regulation are lowest when both the government and industry of one country have opted for the same course of action (e.g. pro-industry) and hit upon countries with the opposite option (e.g. pro-environment). A unilateral course of environmental action may be weakened upon changed domestic policy options, or else interrupted or terminated when government ranks its own economy higher than environmental protection.

The fact that actors are heterogeneous is highly relevant in order to solve problems of cooperation. If one looks at the readiness to take the initiative through pre-commitment or leadership (which is unevenly distributed in the actors' population), one can distinguish three possibilities:

1   An actor takes the leadership role, but remains isolated.
2   An actor takes the leadership role and triggers a bandwagon effect.
3   An actor takes the leadership role but encounters antagonists which leads to a deadlock.

The question about the conditions for successful international cooperation is thus transformed into the question: Under what conditions do actors take the initiative and under what conditions are the others ready to follow?

Networks and systemic variables

It is beyond the scope of this book to address the question whether the model of functional differentiation is a good model to describe the social reality of contemporary societies. A post-Parsonian sociological mainstream seems to favour this model (Alexander 1985; Alexander and Colomy 1990; Collins 1986). According to this, different social realms or spheres can be set off against each other, for example, politics, science, religion, art, economy/industry and the public. Sociological systems theory gives this a special twist by applying autopoietic thinking to it, thus supplementing the principle of functional differentiation with the principle of 'operative closure' (Luhmann 1995). According to this view, societal subsystems are operatively closed; they are not able to give or take input from each other. The advantage of this concept is an analytical precision that might work at the expense of not being able to grasp grey zones of reality (see my own attempts to come to grips with this: Grundmann 1991, 1998). New approaches in the field of science studies claim, for example, that science is undergoing a structural change in which traditional disciplinary boundaries and boundaries between science and non-science are blurred or drawn anew (Gibbons *et al.* 1994; Gieryn 1995; Jasanoff 1995; van den Daele *et al.* 1979; Weingart 1981). Gibbons *et al.* distinguish between two modes of knowledge production which they call mode one and mode two. They see a relative decline in knowledge that has been produced within a (mono-)disciplinary context and hierarchical division of labour (mode one). At the same time, knowledge production that is transdisciplinary and spans several fields of knowledge has increased. More and more knowledge is created in contexts of application. Additionally, non-scientific problem orientation and organisational flexibility play a big role (mode two). The influence of scientific results on the policy process is as remarkable as the influence of social expectations on the research process. Non-scientific actors have a say in what topics should be researched, and where and how to do the research (cf. also Weingart and Stehr 2000). The level of analysis is not at the systemic level but at the level of the actors. Specific actor constellations lead to specific results (Braun 1993; Krohn and Küppers 1989; Teubner 1993; Weyer 1993).

But how would a systemic analysis proceed? Apart from science, the public, industry and politics play an important role. How are the chances of influence distributed between them? There are four main constellations in which different systems control or dominate the process: science, economy (industry), politics and the public (media). One can expect and demand specific things from industry (stopping production), media (setting the agenda) and politics (regulation). From *science* one can only demand what it does in any case, create knowledge. This privilege says little about its political effectiveness, which is contingent upon many different factors. However, in such moments science is quite likely to receive funds. If it gets the monopoly of defining the situation, it can help transform scientific results into political practice; politics would profit from the underlying scientific legitimation (Mukerji 1989). It is a different matter if the public accepts this course of action – there is a likely difference between scientific and lay risk perception (Jungermann *et al.* 1988; Kahneman *et al.* 1982). And a purely scientific legitimation might still not be enough to provide a democratic legitimation. In fact, we are dealing with a tension between competence and legitimacy. As Shils puts it:

> Laymen do turn to scientists for advice that they think is both objective and definitive, but it is often the case that the existing stock of valid knowledge is not sufficient to justify an unambiguous assertion regarding the costs and benefits of a particular policy . . . Objectivity is a very crucial element in the giving of scientific advice. Advisors are too frequently chosen not so much because the legislators and officials want advice as because they want apparently authoritative support for the policies they propose to follow. It is obvious that in complying with these desires, the legislators and the officials are in

collusion with the scientists to exploit the prestige that scientists have acquired for objectivity and disinterestedness.

(Shils 1987: 201)

If one recalls the four types of regulatory policy, one expects that *industry* dominates in cases where costs and benefits are distributed in such a way that it alone would bear the costs of regulation, the public at large would enjoy the benefits but no speakers would turn up. More complicated is the constellation where such speakers exist; in this case one cannot assume that industry will prevail.

A *political* dominance is possible in cases where we have the pattern of a 'majority policy', i.e. benefits and costs are diffuse and it is up to politics to find a majority for regulation. This is likely where policy entrepreneurs (Kingdon 1984) push for a specific solution. This reduces uncertainties of political expectation but requires that scientific results and the public at large tend towards the same direction – a highly improbable constellation. For example, more than half of the British citizens rejected the ban of beef on the bone, introduced in December 1997 and lifted again two years later.

Let us finally consider the dominance of the public. As various studies have shown, mass media has indeed a big influence within society (de Haan 1995; Hannigan 1995; Iyengar 1987; Keplinger 1988; Schulz 1976). Mazur (1981) has argued that an increased media attention on scientific and technical controversies leads to a risk averse reaction from the public, no matter what bias the coverage contains. Lay people seem to take the mere fact of media attention on risk topics as indicative of a serious problem – a position that seems to have a certain rationality (cf. Elster 1979; Føllesdal 1979).

As these short reflections on the potential of one societal subsystem make clear, this is not likely to be located within one of them. Nevertheless, there are several attempts at explaining this case in a monocausal, reductionist manner. As I shall show in Chapter 7, there are mainly economic and cognitive approaches to explain regulations in this case. Interestingly, the variables *public* and *politics* have been largely neglected.

My own approach tries to explore the combinations that make an 'orchestrated' dominance of actors of different subsystems possible. The most promising combination seems to be the one in which actors from different subsystems and organisations network together and propose certain political goals. This leads to a self-reinforcing feedback (mainly by means of research funding and media attention).

The most striking characteristic of these networks is their change in size over time. They gain and lose allies, their resources grow and shrink, actors and resources even change sides. A forecast about the end result a priori is not possible since unexpected resources or actors may crop up at any time. However, this approach allows us to generate the general hypothesis that differences in public credibility will lead to a preliminary decision of the outcome of the controversy.

Method

This book is a case study that deals with the regulation of ozone depleting substances in two countries. Additionally, a third problem, which goes beyond the country comparison, is addressed. This is the emergence of international cooperation to protect the ozone layer. These cases cannot be added up nor can they be reduced – international cooperation is more than just the sum of the national cases. However, international cooperation cannot be understood without the analysis of the national level. In any case, the theoretical design of the case study requires a clear specification of the research questions, the theoretical propositions, of hypothesis building, of connecting hypotheses and data and of interpreting the results. The general questions are: How were CFC emissions reduced on a global scale? Which differences and common features are there between the USA and Germany?

Case study

The case study method is generally preferred when we have more variables than data points (cf. Yin 1994). This method does not allow us to establish explanations in the strict sense, since we are not able to measure causal relationships. The style of presentation is therefore mostly narrative and does not achieve the precision of quantitative data analyses. Theoretical explanations are possible if there is a theoretical framework and the data have been ordered chronologically. Since cause and effect are not reversible in time, this allows for causal inferences. The explanatory power increases if one can refute competing theories. As I shall show, the network approach developed here suggests an explanation of regulations on the national and international level and does better than other approaches.

A problem with case studies is how to avoid false generalisations, although generalisations in case studies are not impossible in principle. Yin distinguishes between 'statistical' and 'analytical' generalisations. The latter is possible where data support a theory. If two or more cases support a theory, one may speak of replication. This is not to be confused with the enlargement of a sample in statistical tests (for this reason it is not a 'statistical' generalisation). The performance of several case studies is preferable to replicating an experiment.

Time frame and scope

The time span covered in this book extends from the publication of the CFC-Ozone-Hypothesis in 1974 to the international treaties of Montreal (1987), London (1990) and Copenhagen (1992), with some glimpses beyond. This is a 20-year period, which seems to be natural. However, it goes to the very limits one researcher can master in a reasonable time, provided one seeks historical detail. Contrary to many social science publications on this case which have appeared recently, I do not limit myself to the analysis of the international regime for the protection of the ozone layer. I analyse the first period of regulation on the national level and the comparison of two countries. In so doing, I attempt to put the research onto a broader footing. The first point indicates breaks and continuities between both periods, the second to differences and common features in both countries.

Sources

I have used the following sources:

- relevant scientific publications in the field of the atmospheric sciences;
- official reports of governments and parliaments (or such which have been commissioned by them, like the reports by NASA and Enquetekommission);
- publications of international organisations (World Meteorological Organisation [WMO], UNEP);
- press reports;
- social science secondary literature on the ozone case;
- archives, especially of the German Umweltbundesamt (UBA), Paul Crutzen and F. Sherwood Rowland;
- interviews with experts, conducted by the author.

Interviewing took place between September 1994 and June 1995. The selection of experts (N = 52) was done on the basis of the secondary literature and by snowball sampling. The interviews lasted for between 30 minutes and three and a half hours. Some experts were contacted more than once; on the whole, they were willing to answer questions later on. The experts can be grouped into atmospheric scientists, representatives of the chemical industry or their interest organisations, environmental organisations and members of parliament, public administration or diplomatic service. I used a semi-structured questionnaire. The largest group of interviewees comes from science (N = 27), of which the main share are scientists working in the USA (N = 16).[22] Other experts are with industry (N = 5), politics, diplomacy and administration (N = 20). This classification reflects their function during the interviewing period, which is not in all cases identical with their present function. The interviews were transcribed and analysed with the help of the software package ISYS. Data have been anonymised.[23] However, many of the interviewees are named in the literature and in press reports (and did not insist on their anonymity). Three of the interviewed scientists were awarded the Nobel Prize for chemistry in 1995. The strict observance of the anonymity principle would have left informed readers bewildered. For this reason, I have at times named central actors, citing an accessible source where possible.

2 Ozone science

Before we examine the question of how the ozone layer became perceived as endangered, some background information seems to be useful. Based on sources like Dotto and Schiff (1978), Graedel and Crutzen (1995) and my interview material, key scientific concepts and a common body of knowledge are introduced which then became challenged. The chapter traces major developments in scientific understanding and practices, focusing on scientific specialities and interdisciplinarity as well as on scientists' normative orientations and belief systems. It concludes with a short review of the revisionist 'backlash' against the emerging consensus view of the late 1980s.

The ozone layer

At the beginning of the twentieth century French scientists discovered the stratosphere. This is the region above the troposphere where temperature is not decreasing with increasing height – as is the case with the troposphere – but is actually increasing. The temperature increase starts at about 12 km and continues up to about 50 km, where it reaches values like on the earth's surface.

This important difference between the troposphere and the stratosphere is explained by the creation and destruction of ozone in the stratosphere, a process during which solar ultraviolet energy is converted into heat. Ozone concentrations are maintained by mechanisms of positive and negative feedback. In 1930, Sidney Chapman, the founding father of aeronomy, suggested two chemical reactions as main mechanisms. In the first an oxygen molecule (O_2) is split by UV-light into two oxygen atoms. In the second, an oxygen atom combines with an oxygen molecule and forms ozone (O_3).

$$(UV) + O_2 \rightarrow 2\,O \tag{1}$$

$$O + O_2 \rightarrow O_3 \tag{2}$$

The creation of ozone thus requires only two ingredients: UV-light and oxygen. The further one ascends in the atmosphere, the more UV-radiation increases, and the more oxygen pressure decreases. Therefore, one would expect a point at which ozone creation reaches its maximum. If this were everything, all oxygen would be transformed into ozone, a positive feedback loop that would come to an end only when there was no oxygen left. However, Chapman also identified a complementary negative feedback loop. Here, the free oxygen atoms destroy an ozone molecule by transforming it back into two oxygen molecules.

$$O + O_3 \rightarrow 2\,O_2 \tag{3}$$

Since the speed of ozone destruction is directly proportional to the available ozone, the more ozone that is available, the faster it will be destroyed. Chapman assumed that the conditions in the stratosphere would lead to a balance between ozone creation and destruction. Such was the conventional wisdom for a long time. The chemistry of the stratosphere was established and dominated for many years by Chapman's pioneering work. No one saw any theoretical problems to solve. Researchers thought that ozone chemistry was well understood; there were some indications of anomalies but they were pushed aside:

> To prove that there was no problem, they simply took those kineticist data of lab simulations which were in accordance with the old theory – but which were false. They were content, they thought: we have solved the problem, now we embark on something new.
>
> (GEAS 25)

The rethinking about ozone in the stratosphere really started in the mid-1960s, and centred around possible reactions of species related to water – such as hydroxyl (OH) and HO_2. However, to get a balance scientists had to make up rate constants for reactions that had not yet been measured. When they were measured, the problems emerged. Observations and measurements showed that ozone concentration was much lower than would be expected according to Chapman's theory. Therefore, scientists began to assume the presence of other ozone depleting mechanisms that were as yet unknown. Which substances could deplete ozone? The main composites of air, oxygen, nitrogen, carbon dioxide and water vapour could be excluded since they do not react with ozone.[1] Only a remote possibility pointing to trace gases remained. Since trace gases were found in concentrations well below the tiny concentrations of ozone, it was far from clear how they could possibly destroy ozone. The answer was to be found in catalytic chain reactions whereby one substance destroys another without destroying itself. This was the first real theoretical innovation in the field of atmospheric chemistry after Chapman. Laboratory experiments had shown that nitrous oxides (NO_x) and hydrogen oxides (HO_x) were effective catalysts. In the 1950s, various researchers suggested plausible ways that explained how these substances could form in the stratosphere. However, they did not find the 'missing' ozone destruction process. This was done by Hampson and Hunt for HO_x and by Crutzen for NO_x (Crutzen 1970). In 1974, Molina and Rowland put forward the hypothesis that chlorine was an even more effective catalyst. The following equations show the catalytic process of ozone destruction (the catalyst is shown as M, cf. Graedel and Crutzen 1995).

$$M + O_3 \rightarrow MO + O_2 \tag{4}$$

$$O + MO \rightarrow M + O_2 \tag{5}$$

$$\text{Net: } O + O_3 \rightarrow 2\,O_2 \tag{6}$$

As can be seen, there is a net loss of ozone: the catalytic chain transforms one oxygen atom and one ozone molecule into two oxygen molecules.

In the 1960s, this work became relevant as plans were made to build a fleet of supersonic aircraft in the USA. In one of the first technology assessment projects about the effects of such a fleet, the role of nitrous oxides and their effect on ozone was thematised (Crutzen 1970; Johnston 1971). Some years later, chlorine became another substance that was thought to play a role in the destruction of stratospheric ozone. The issue broke at a scientific conference in Kyoto, Japan when US-researcher Richard Stolarski presented findings he had made, together with his colleague Ralph Cicerone, in the context of yet another big technology project: NASA's planned *space shuttle*. They calculated that the rocket would emit large amounts of hydrochloride (HCl) and therefore inject considerable quantities of chlorine into the stratosphere. The estimated amounts were dependent upon the number of yearly flights. Based on fifty annual take-offs, they arrived at a figure of 5500t HCl. However, Stolarski did not mention the *space shuttle* during the conference (Dotto and Schiff 1978: 123). As potential source for chlorine he suggested volcanic activity.

The initial hypothesis

In 1974, Rowland and Molina pointed out that chlorine from CFCs is a potent ozone destroyer. Measurements of CFC concentrations around the world (Lovelock was the first to carry them out with considerable accuracy) had shown that world CFC production was consistent with the atmospheric concentration of CFCs; in other words, these substances do not dissolve but accumulate and diffuse through the atmosphere. Their lifetime was estimated to be between 40 and 150 years. The upper atmosphere (>24 km) was assumed to be the only sink where CFC are photolysed by UV-light. As with NO_x, a catalytic chain reaction occurs, in this case a reaction of active chlorine with ozone, only the chlorine is much more effective. The authors highlight the fact that the atmosphere's capacity to absorb chlorine is limited, and that serious consequences could follow. Rowland and Molina's article ends with the warning that environmental problems will persist for a long period after the reduction of emissions. They claimed a causal chain with the following components: CFCs reach the stratosphere intact; there they are split by UV-light; chlorine radicals emerge; these destroy ozone molecules in a catalytic chain reaction (including atomic oxygen in the upper part of the stratosphere). The expected ozone reduction is estimated to be between 7 and 13 per cent in 100 years. These figures were dramatic, based on the conventional estimate that 1 per cent ozone loss leads to an increase in UV-B radiation of 2 per cent, which in turn was assumed to cause 2 to 3 per cent more cases of skin cancer and eye cataracts, apart from other effects on the human immune system and plant and sea life.

This was bad news for CFC manufacturers. They attempted to track down all the weak points of the hypothesis in order to exonerate CFCs. Logically, every possibility that makes ozone depletion seem impossible or insignificant is suitable to this purpose. This is the case if only small quantities of CFCs reach the atmosphere, or natural sources exist that are greater than the anthropogenic sources; if the lifetime of CFCs is relatively short; if CFCs disappear in tropospheric sinks before they can climb into the stratosphere; if the kinetic reaction rates of important molecules that play a role in the depletion of stratospheric ozone proceed slowly; if no measurable ozone loss occurs; if inconsistencies appear in the theory, or between theory and measurements. If a reduction in ozone actually had to be dealt with, there was an additional method of defusing the problem. This is to use the possibility that no immediate or demonstrable connection exists between ozone reduction and possible effects (i.e. frequency of cancer, plant growth, eye diseases, etc.). In attacking the Molina–Rowland hypothesis, industry made use of all these strategies (see Chapter 4).

After the publication of Molina and Rowland's hypothesis it became clear that CFCs constituted an industrial source for chlorine much bigger than volcanoes or space shuttles. In the ensuing years, important scientific research was carried out. Model calculations were based on CFC emissions, their average lifetime, and the probable chemical reactions in the atmosphere. The models were one-dimensional (which I explain in the next section) and used only gas phase reactions. All controversies during this first decade (1974–84) revolved around predictions of future ozone depletion calculated from these variables.

The research field made its next important progress after the discovery of the Antarctic ozone hole in 1985. The explanation of this phenomenon revolutionised the understanding of the atmosphere again. Early models calculated the main ozone loss in the upper stratosphere (*c.* 35 km) and were based on homogeneous chemical reactions, as sketched above in equations (4)–(6). However, the occurrence of the Antarctic phenomenon could not be explained this way, mainly because there is too little light during the polar night (July–September) and too little atomic oxygen.[2] Those massive ozone losses required another explanation. The dominant explanation today holds that during the Antarctic winter, with its extremely low temperatures (minus 80 degrees Celsius), ice particles made up of water, sulphuric acid and nitric acid form. As soon as the first rays of sunlight come up in the Antarctic spring (late September), there are heterogeneous reactions – reactions between substances in different states – on the surface of these ice crystals involving mainly $ClONO_2$ and HCl. These gases usually function as reservoir gases, since they bind chlorine. However, with the upcoming sunlight individual chlorine atoms are set free.[3] Without going into further detail here, the process is terminated by the combination of one ClO with another ClO, only to be started anew when sunlight once again releases atomic chlorine from the product.[4] However, it is not clear if this mechanism can explain all of the ozone losses in middle latitudes. As one scientist put it,

The difficulty is that at low altitudes in mid-latitudes it's not just chemistry that controls ozone but motions, and unless you have good simulations on these motions on a year-to-year basis, you're not quite sure what the cause is. That's still a hot interesting scientific topic of current research.

(USAS 30)

However, the results of polar ozone research suggest that the main losses of ozone are in the 12–22 km region and not, as previously assumed, above 35 km.

Mid-latitude ozone loss isn't as clearly understood as polar ozone loss. This is a big signal, a big process, a big change, and it's localised . . . Mid-latitude chemistry is much subtler. What's going on here is a subtle balance shift, not a big change. You get 100 per cent ozone loss in the Antarctic, that's not hard to miss. In mid-latitudes we are talking about 5 per cent, 10 per cent, small things.

(USAS 17)

It is not clear if decreasing ozone concentrations in middle latitudes are caused locally or if they are caused by influx of ozone poor (or chlorine rich) air from the poles. Local explanations assume that aerosol particles (such as sulphur particles emitted by volcanic activity) provide a similar surface for heterogeneous reactions as ice particles in Antarctica. Another explanation assumes iodine as a further catalytic substance, besides chlorine and bromine.

One thing that is not explained is: Why do we have so much ozone depletion in the mid-latitudes in the Northern hemisphere? All over the 10 years we had 5 or 6 per cent depletion, and the models cannot simulate that. So is it because there are leaks from the vortex in the Arctic? Some people say yes, some say no. Or is it a local depletion? If so, why? Some people claim that aerosols played a big role. Maybe. There is a new idea that not only chlorine and bromine but also iodine plays a role. If so, how does it get there? It has a lifetime of only 3 to 5 days, so it doesn't have time to get into the stratosphere except if there is a strong convection process, like thunderstorms.

(USAS 41)

This was the state of knowledge at the time in 1994 to 1995 when I conducted my interviews; since then, no radical changes have occurred (cf. Walker 2000). It is clear that science has not achieved an encompassing consensus about ozone depletion in the stratosphere (Haas 1992b). But equally problematic is the statement that this problem is totally undecided (McInnis 1992). There is a consensus about the *explanation* for the Antarctic ozone hole – although even here not all the questions are answered – and about the *observation* that a global decline in ozone concentrations has occurred. I shall return to both points in greater detail.

Growth of the field

Trying to map the development of the field of ozone research with sociometric methods, I chose four key words in the *Science Citation Index* (Figure 2.1): *stratospheric ozone, ozone depletion, atmospheric chemistry* and *halocarbons/chlorofluorocarbons*. The most striking finding is that there is a sharp increase after 1989.[5] This finding alone suggests that major scientific questions were not resolved at the end of the 1980s.

There are three distinct practical fields within the atmospheric sciences: modelling, field experimentation and laboratory experimentation. Laboratory work consists mainly in identifying important chemical reactions and determining rate constants. In the atmosphere the presence of chemical species is measured (*in situ* measurements), and researchers make observations on ground stations. The relative prestige of these groups shifts in the course of time. It can be said that it was a conflictual process but ultimately led to closer cooperation.

Modellers have set themselves a daunting task in creating mathematical models of the atmosphere. These models can be distinguished into zero-dimensional, one-dimensional, two-dimensional and three-dimensional models. Zero-dimensional models only contain chemical reactions without transport. A one-dimensional model has only one variable, altitude, and assumes average values for latitude, usually 30 degrees (either North or South) and longitude. Two-dimensional models add latitude as a variable, and three-dimensional models add longitude. The computational power required for each added dimension goes up enormously, and the mainframe computers of 1974 were about as powerful as the current hand calculators for high school students. (Some scientists carried a slide-rule to do numerical calculations until mid-1973.) The move to 2-D in the early

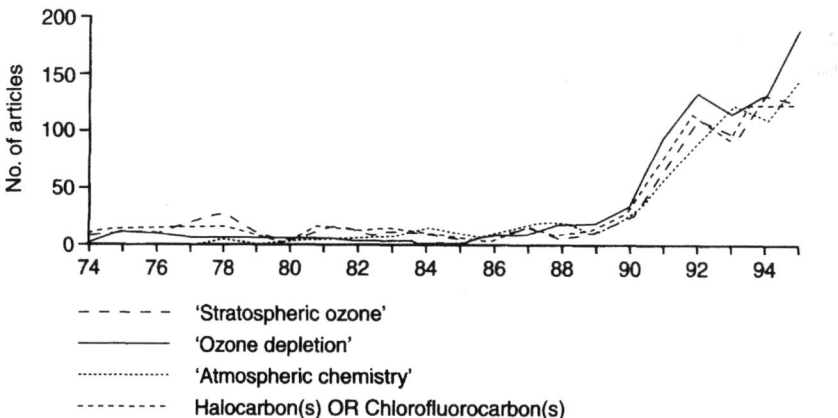

Figure 2.1 The expansion of ozone research, selected key words, citation frequency per year

Source: SCI, own calculations.

1980s was facilitated by the availability of far more computational power. As one modeller put it:

> the more complicated a model becomes, 0-D, 1-D, 2-D, 3-D, the more uncertainties you have, you are creating more sources of errors. My own experience, having all these models, and asking: Which is the most appropriate model? The answer is: It depends on which problem you are treating.
>
> (USAS 41)

Good modellers are able to balance the advantages and disadvantages and thus reach a good compromise between these types of models according to the problem they want to investigate. Simple models cannot claim to represent the atmosphere because of their simplicity. Increasing complexity does not help very much since an increase in dimensions also leads to an increase in potential sources of error: 'Models are always a simplification of reality and in fact, what we say, is: junk in, junk out. Whatever you put as hypothesis in your model will be reflected in the results of your model' (USAS 41).

At the beginning of the 1980s, when modellers started to experiment with two-dimensional models, ozone depletion was estimated to be very low for the next 100 years or so. Some even saw an increase of ozone. The main reason for these estimates was that the models only calculated net changes in the ozone column, but did not consider the changes that would result from a change in the temperature profile of the atmosphere. In fact, the net change may amount to zero if ozone is destroyed in the upper stratosphere since relatively more ozone is created in the lower stratosphere. Two reasons account for this effect: first, ozone concentration decreases with increasing altitude. Oxygen concentrations are higher in the lower stratosphere and therefore the probability that UV-radiation creates more ozone is greater (since it hits upon more oxygen molecules). Second, because of a reduced ozone layer in the upper stratosphere, more UV-radiation hits on the oxygen molecules further down, leading to a negative feedback loop. In the 1970s, this had been interpreted as a kind of self-healing mechanism of nature.

Modellers now make the change from two-dimensional to three-dimensional models which comprise complex chemistry. These models are very expensive and require an enormous amount of calculation time. The effort is only justified in clearly defined cases.

> 3-D models are very expensive. And they are mostly focusing on transport with as much chemistry as you can afford. My 3-D model has 150,000 points. So you have to solve your chemistry 150,000 times at each time step. While in the 0-D model only once. If you compare the 2-D and 3-D models in the ozone case, and you have in the 2-D model too little ozone at 40 km, the shape not being perfect, maybe too much ozone in the lower stratosphere, then you will have the same problem in the 3-D model. You haven't solved anything.
>
> (USAS 41)

Cooperation between modellers and experimenters seems functionally required but did not grow naturally. The following points make cooperation look very reasonable. Modellers and experimenters trying to synthesise the results of the field fail when working in isolation from each other. Experimenters who always analyse one partial detail of the larger context try to cover areas that are outside their expertise. The modellers try to put together the puzzles provided by the experimenters, thereby 'expropriating' them. Envy and tensions are the result. In some cases the solution is to publish research findings in co-authorship. There are attempts by experimenters to theorise their own findings that are regarded as ridiculous by their modeller colleagues ('But you also have some cases where the experimentalists try to make their own explanations and they are really bizarre', USAS 17). However, there are also scientists who started out in either speciality and gained a grasp of the whole field:

> Sherry Rowland is not a modeller. He is someone who understands fundamental mechanisms and can separate the rate-limiting step for a whole lot of chemical garbage, and so the fact that this photochemical system could produce chlorine monoxide (which was the rate-limiting catalytic agent) was the key question.
>
> (USAS 8)

In the judgement of atmospheric scientists, the modellers are seen as having played both a positive and a negative role. Positive is the fact that they postulated the presence of certain chemical species in the atmosphere that had not yet been observed. This judgement is made by modellers and experimenters alike.

> The models did not give the right answer. Progress was made through lab experiments but also through the modellers who postulated substances like chlorine nitrate on the basis of their theories, although no one had actually measured it. To be sure, the reaction rates were known; the modellers then inferred that there must be measurable quantities in the atmosphere which were measured later on.
>
> (GEAS 35)

Their role was judged differently when their one-dimensional models, with their unrealistic assumptions, were trusted more than empirical observations. In a rather dramatic way the ozone hole was 'missed' by NASA since it had programmed an ozone measuring instrument aboard their satellite in a way that excluded abnormally low values.

The influence of modellers and theoreticians seems to be the same as in other disciplines: they are at the top of their discipline.[6] It is their models that bring together the many various details into one all-encompassing picture. However, the models have also been used as a basis for policy decisions, although they did not predict the ozone hole and could not explain it for two years after its discovery.

Experimenters are sceptical as regards the explanatory value of those models. They trust them only if the most important data with which the modellers work have been verified experimentally. For a long time it looked as if some important model predictions did not concur with the experimental evidence.

> In fact we were involved in a lot of controversy essentially pointing out that a lot of the models which we used to predict ozone loss rates were in direct conflict with observations. . . . And we went through National Academy Report after National Academy Report where the models were not calculating those rate-limiting steps in a proper way, so they couldn't possibly represent the natural system. . . . But that didn't stop the predictions from being published year in year out.
>
> (USAS 8)

Models played an important role when testing hypotheses about presumed chemical and dynamical processes in the stratosphere. In other words, knowledge about the atmosphere is also achieved by means of model calculations. Input data are made up of initial values, postulated and known chemical reactions and diffusion parameters. In cases where the model simulation yields results that are consistent with measurements, the explanations are regarded as plausible. Such testing of models with reality has only started in the recent past. In 1987, leading modellers wrote in a scenario paper for UNEP: 'No model has yet been adequately validated against the real atmosphere (e.g. current ozone distribution) and their reliability for predicting future states of the atmosphere is still uncertain' (UNEP 1987: 3).

The ozone hole

In May 1985 scientists of the British Antarctic Survey (BAS) published an article in *Nature* in which they claimed to have found abnormally low ozone concentrations over Halley Bay, near the South Pole (Farman *et al.* 1985). The article contained a very suggestive graph that plotted the decreasing ozone levels over springtime Antarctica together with increasing CFC levels in the Southern hemisphere (see Figure 2.2). The BAS measurements were single-location readings taking the ozone concentrations above the Halley Bay station from 1958 to 1984, using a Dobson spectrometer. Many researchers questioned whether a dramatic conclusion could justifiably be drawn from these readings. At all events, NASA's more up-to-date and far more expensive satellite instruments for measuring global ozone values failed to provide any confirmation of the BAS readings at that time. In addition, as the BAS was a group unknown to the core of the international research community, many scientists initially considered the data supplied by the BAS to be nothing more than instrument errors.

After checking its satellite data, NASA eventually confirmed the alarming readings (Stolarski *et al.* 1986). The information was well known in the community by late August 1985. They had thousands of low ozone data points every day within

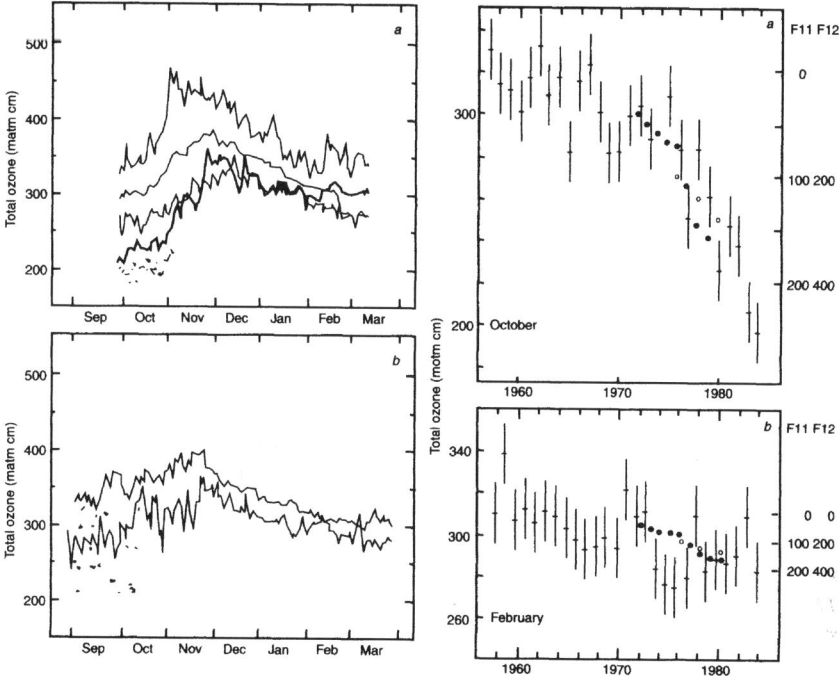

Fig 1 Daily values of O₃. *a*, Halley Bay: thin lines, mean and extreme values for 16 seasons, 1957–73; thick line, mean values for four seasons, 1980–84; +, values for October 1984. Observing season: 1 October to 13 March. *b*, Argentine Islands: as for Halley Bay, but extreme values for 1957–73 omitted. Observing season: 1 September to 31 March.

Fig 2 Monthly means of total O₃ at Halley Bay and Southern Hemisphere measurements of F-11 (•, p.p.t.v. (parts per thousand by volume) CFCl₃) and F-12 (o, p.p.t.v. CF₂Cl₂). *a*, October, 1957–84. *b*, February, 1958–84. Note that F-11 and F-12 amounts increase down the figure.

Figure 2.2 Farman *et al.* 1985. Monthly means of ozone and CFC 11 and 12 over Halley Bay in October and February, 1958–1984. Note the last sentence in the legend: F11 and F12 concentrations increase down the figure

Source : *Nature* 315, 16 May 1985. Permission by Macmillan Magazines Ltd.

the polar vortex. Using a special graphical representation of the NASA data (distribution of the global concentration of ozone), it was possible to visualise the geographical extent of the destruction of Antarctic ozone. This was a convincing visual demonstration to the onlookers at conferences and in the policy game. So the phenomenon was real, but what caused it? Were CFCs implied or was this a natural phenomenon that had gone unnoticed so far?

Dynamicists thought that the ozone hole was caused by the specific meteorological conditions present over the South Pole during the Antarctic winter. These conditions include a stable polar vortex, in which huge ozone losses are accompanied by a drastic fall in temperature. Dynamicists argued that this was the reason that ozone rich air outside the vortex would not be able to replenish ozone poor air within the vortex. A dynamicist explained the problem as follows:

We only had two pieces of information: ozone is going down and the temperature is going down. We didn't know anything about the chemistry at the time. So in '85–'86 we only had these two pieces of information from satellites. And there were two hypotheses to explain it: one of which is, the air is coming up adiabatically and it's cooling as it rises and it's pushing the ozone layer up at the same time. So there could be some dynamic phenomenon developing in the Southern hemisphere which is creating this. The other hypothesis was: if the ozone is going down chemically, then you have less heating of the stratosphere and it will get colder. So it's a chicken–egg question. It seemed to me that it was very premature to assume that it was chemical.

(USAS 17)

This provoked the chemists who did not see any support for a dynamic explanation. One of the chemists told me,

Personally I just get a little nervous when people say 'It must be dynamics!' without any reason. If you can observe something, like temperature or wind changes, then it would be different. But just throwing up the hands and saying It must be dynamics. . . . Why?

(USAS 36)

It is hardly surprising that dynamicists would reproach the chemists in exactly the same way:

I was very much against people saying: 'It's got to be chemical'. Sherry Rowland said: We can see that from the planet Mars, 'It's got to be chemical'. How does he know? Before we jumped into regulating (people lose their job if you regulate!) I thought we needed more research. I was very upset with the chemists, because I thought they were making a big assumption.

(USAS 17)

The whole field at that time [at the beginning of the 1980s] was dominated by chemists. The original theory was a chemical theory and they deal with an atmosphere which is more or less static, they used to have 1-D models for vertical profiles. Dynamicists got into this field late. Physically-based foundations were not available in a manageable model. The 3-D models often had . . . primitive chemistry in it. We had two extremes: chemists having very sophisticated chemical schemes but not transport and the dynamicists who . . . could not incorporate the complicated [chemistry].

(USAS 38)

Through a series of experiments, both in the laboratory and in the field, the dynamicists were eventually persuaded that dynamic factors would not suffice to produce the ozone hole. There had to be chemical reactions, among them chlorine as major substance (Shell 1987). This consensus that emerged around 1988 was based on a period of intense discussion and controversy. The participants of this process learned much of each other and then applied their knowledge to the ozone layer in the middle latitudes and in the Arctic. The losing side in the controversy has accepted its defeat. One of them said: 'So we were disappointed that we were wrong but happy of having been able to inject good science into the process. Hypotheses are meant to be rejected' (USAS 11). This, in fact, sounds like the ideal of Popperian science. However, we shall have the opportunity to further analyse the mundane practices of scientific work and see that scientists have not always conformed to this ideal view of science.

Dynamics is now regarded as much more important than in previous years, although it could not explain the ozone hole. Today all good models have a combination of dynamics and chemistry. A dynamicist put it this way:

> Today we all get together and don't talk about cultural gaps any more. . . . When we talk about 'potential vorticity dynamics' the chemists no longer panic. The fields have really been culturally merged in such a way that we pretty much speak the same language. That's very exciting.
>
> (USAS 11)

Today, chemical explanations all make use of 'heterogeneous reactions' which are regarded as the centrepiece of the ozone destruction mechanism. This type of chemical reaction was a new discovery in the field of ozone research. Before 1986, hardly anyone considered this theoretical possibility to be real. After the discovery of the ozone hole, the scientific community rather haphazardly tried to come up with an explanation quickly.[7] However, for the international political negotiation process leading up to Montreal, a commonly accepted explanation was not available.

> In 1985 we were saying to ourselves that making these speculations, we don't know when we will be proven wrong. Maybe in 10 years, that was the feeling. It is remarkable that the scientific community mobilised and got these expeditions to this remote place of the planet to actually find out what the answer is. Of course there is a lot of excitement associated with it. It is very rare for chemists to see a new chemical reaction at work producing such dramatic effects. It's an opportunity of a lifetime. That's why there is so much enthusiasm for going down there.
>
> (USAS 38)[8]

The core group of atmospheric scientists came together at the beginning of May 1988 in Snowmass, Colorado to discuss the results of two expeditions to Antarctica. Before the meeting started, it seemed as if neither dynamicists nor chemists were ready to give in. The conference chairman used a metaphor from the old days of American railroads to express what he felt the meeting was going to be:

It's like the old days of railroading when two trains, unable to communicate, would speed unknowingly toward each other on the same track through the middle of Kansas. Eventually, the was going to be a cornfield meet between the two – guaranteed to be messy.

(cit. in Nance 1991: 183)

To be sure, both camps collided (in a five-day conference with 69 papers), but in the end there was a common understanding emerging.[9] The dynamicists realised that the exceptional meteorological conditions over Antarctica were not sufficient to make the ozone hole. A minority among them held the belief that the reverse was also true, that increased chlorine concentrations would not be sufficient to produce an ozone hole. The final evaluation of the data of the second expedition to Antarctica (1987) led to a refutation of the dynamicists' hypothesis.

Interdisciplinarity

It has been noted that in contemporary societies powerful tendencies of differentiation are at work. These also act on the production of knowledge, which, as a result, becomes ever more fragmented (Dickens 1996). How can one imagine an integrated body of knowledge that comprises several scientific disciplines? To expect a new unity of science is probably utopian. However, problem-orientated research involving several disciplines and areas of knowledge could lead to some degree of merging specialised fields. This is by no means an automatic process. An increased number of specialities will only increase the likelihood of re-combinations, not the actual occurrence. While multidisciplinarity joins together different bodies of knowledge in an additive way, interdisciplinarity leads to an interpenetration of fields of knowledge and thus to innovation (Friedman and Friedman 1990).

Looking at the field of atmospheric sciences, one sees a two-fold process of cooperation emerging: between modellers and experimenters, and between atmospheric chemists and dynamic meteorologists. This seems to be rather surprising when we consider that incentives and rewards are usually based on disciplinary rather than interdisciplinary standards and that scientists are trained in one specialty rather than in a broader area. It is a commonplace to say that our academic institutions seem to favour extreme specialisation (Klein and Porter 1990). So, which factors can account for the emergence of cooperation in this case? I can see two: special skills of researchers and the existence of local research groups. I address each in turn.

Scientists with many skills

There is a widespread scepticism about the prospects of interdisciplinary research in general and specifically about environmental research. True, most inter-disciplinary initiatives do not reach beyond 'aggregating mono-disciplinary achievements and competencies': 'Either existing knowledge . . . is applied, or the cumulative theory building continues within the boundaries of a single discipline' (Weingart 1974: 24). There is the need for a common language that need not be identical with the disciplinary terminology of a narrow speciality. Often this is done by translating the problem into everyday language and back into the disciplinary language, which is very time-consuming (Kaufmann 1987). In sum, an interdisciplinary orientation clearly poses a risky investment for scientists and is only taken if special incentives are offered (Weingart 1987; Weingart and Stehr 2000).

So far the literature has paid little attention to the possibility that interdisciplinarity can grow from below if scientists with manifold competencies and skills occupy a key role (for an exception, see Hollingsworth 2000). It has been noted that 'academic intellectuals', people with an eye towards questions of personal and societal importance, would be the 'ideal personnel' to study cross-cutting issues (cf. Klein and Porter 1990). In the case of ozone some of the main protagonists of the early scientific discussion were outsiders in the field of atmospheric research. However, as individual researchers with a wide horizon and a propensity to acquire competencies in other specialities, they were crucial for the emergence of interdisciplinarity (see Ben-David (1960) for his concept of 'role-hybrids').

Cicerone, Stolarski, Rowland and Molina were not trained as stratospheric chemists. Cicerone and Stolarski were not even chemists; Cicerone had a degree in electrical engineering, Stolarski in physics. They were both interested in the ionosphere, the region above the stratosphere. When they started their research, the atmosphere of the earth was divided into different floors with different institutional responsibilities – for each floor there was one sub-discipline. All of them defended their domain eagerly. Before the 1960s the stratosphere was largely ignored (Dotto and Schiff 1978: 121 f.). Jokingly, it thus was labelled 'ignorosphere' (Crutzen 1996).

Compared with 'normal' projects, interdisciplinary projects have greater difficulty in getting funding. On the part of the referees in the different disciplines there exist various prejudices against bidders from 'outside' disciplines; basically they try to defend their own turf (Porter and Rossini 1985). However, it is also possible that scientists who are more open towards interdisciplinary research and act as referees themselves turn this bias around and favour projects that are also interdisciplinary. My interview material confirms that atmospheric researchers view the interdisciplinary opening in their field very positively.

The main opponents in the early days of the controversy were mainly experimenters with a large knowledge of the whole field and with manifold skills. The following examples show how broad the skills and interests of the main protagonists have been. Lovelock built the *electron capture detector*, conducted field and laboratory measurements and developed his Gaia philosophy (Lovelock 1988). He embodies the innovative and 'entrepreneurial' type of scientist. This often led to clashes with the academic establishment.[10] In the late 1960s, he initiated interest in CFCs when researching the causes of smog in western Ireland.

He used CFCs as a tracer, trying to show that polluted air was drifting from the Continent. In fact, the result of his research confirmed his hypothesis and he published two articles on it. After that, he tried to obtain research grants in order to do similar projects aboard a ship travelling to Antarctica. These proposals were turned down since Lovelock's claims of measuring air concentrations of one part in a trillion were regarded as frivolous.[11]

> One referee said it will be exceedingly difficult at the parts per million level in the air and yet the claimant is claiming to measure them in parts per trillion. It was considered a bogus application and the board felt that its time should not be wasted with frivolous grant requests.
>
> (Interview with Lovelock)

Lovelock's colleagues emphasise his exceptional skills in building instruments. Far from being fraudulent, he pioneered the measurement of trace gases. However, his Gaia philosophy, which bears an immediate relationship to the ozone controversy – a complex system of negative and positive feedback loops that makes nature robust – is viewed with mixed feelings by his colleagues. Some scientists think it is religious speculation, some think it is interesting material for discussion. His early statement that CFCs would pose 'no conceivable environmental hazard' (Lovelock *et al.* 1973: 194, though by 1974 he characterised this as an 'unwise comment') brought him to the pro-industry position in the beginning of the controversy. Today, he thinks that this phrase was the biggest mistake he ever uttered.

Rowland, a radio chemist by training, came to the field of atmospheric chemistry in 1973. He did lab measurements, started field measurements after 1977, and calculated with his erstwhile assistant Molina the hypothesised ozone depletion rates with a one-dimensional model. The lifetime of CFCs was decisive:

> The calculation of the average lifetime of a molecule – the early stage of our work in late 1973 – involved a simple one-dimensional vertical model of the atmosphere: the CFC compounds were subject to two effects, the motion up and down through the atmosphere (put in with an 'eddy diffusion' constant at each altitude), and a rate of photolysis at each altitude, causing it to disappear. We used all of the eddy diffusion profiles worked out by the other scientists who had developed computer models in connection with the work on the SST problem, and showed that all of them gave about the same result, e.g. 40–80 year lifetime for CCl_3F [CFC 11, R.G.].
>
> (Personal communication with Rowland)[12]

At the beginning of the 1980s, Rowland suspected that there were more CFCs in the atmosphere than official figures suggested. He did some field research and published the results in 1982 showing that, in fact, there was more CFC 12 in the atmosphere than had been officially produced. Industry explained the difference with leaks in the production plants. Rowland himself carried out measurements downwind of a big CFC plant which did not confirm industry's claim.

> We then tried to verify this 1.5 per cent leakage, which was a major source. Any big firm should be a major source. And so we got samples downwind from the Du Pont plant in Texas. There was a lot of methyl bromide, carbontetra-chloride, but very little CFC 12 or 11 [on that day]. I didn't believe their explanations, I felt all along that some of their companies were underestimating.
>
> (Personal communication with Rowland)

Rowland combines such different skills as lab kinetics, model calculations and *in situ* measurements. He developed some of these skills during the controversy, driven by the motivation to endure in the struggle with Du Pont.

Like Lovelock, Jim Anderson is a scientist with great skills in the construction of scientific apparatus. In the 1970s, many measurements of chemical key substances came from him. 'Without Anderson's *in situ* measurements the modellers would not have real numbers to put into their models to see if they could reproduce them' (UNAS 2). He built the instrument that recorded decisive measurements aboard a special aircraft during an expedition to Antarctica in 1987. He developed this skill over a long time; precursors of the instrument had been used during balloon experiments in the 1970s. His instrument was very accurate and gave a very clear signal. His results were used after the campaign to close the controversy about the causes of the Antarctic ozone hole.

Paul Crutzen is a former engineer and meteorologist, and a self-taught chemist. In 1968 he did his Ph.D. in meteorology ('Determination of parameters appearing in the "dry" and the "wet" photochemical theories for ozone in the stratosphere') with Bert Bolin in Stockholm and in 1973 completed one of the last higher doctorates (habilitation) in Sweden ('On the photochemistry of ozone in the stratosphere and troposphere and pollution of the stratosphere by high-flying aircraft'). In 1970 he discovered that nitrous oxides could attack the ozone layer. This was the first scientific revolution in the field of stratospheric chemistry after four decades of theoretical stagnation. Crutzen is a modeller who is very interested in man-made changes in global climate and the ozone layer. On several occasions, like Rowland, he took an advocacy position and spoke out publicly. He designed various scenarios about possible effects of an atomic war ('nuclear winter') and the 1991 Gulf War. Crutzen was not only engineer but after 1977 also worked within NCAR's administration in Boulder. During this time he continued his scientific work and initiated an interdisciplinary research project to investigate the relation between the atmosphere and the biosphere (cf. Crutzen 1996).

In all these cases it is remarkable that innovative impulses came from the fringes of established science. Sometimes they were outsiders, newcomers or scientists who dared to go beyond their specialities or who adapted the practical skills they bring to bear onto their academic work. This is surely a risky way to earn scientific reputation. To put it in a slightly different way: scientific business as usual would have delayed many of the discoveries, or the discoveries would not have been made at all. The researchers have taken a high risk for their own career.

Localities of research

Many key papers in the field have been co-authored by two researchers who work together in close proximity either in the same faculty, the same institute or in a nearby institution. Rowland and Molina, for example, cooperated with modeller Donald Wuebbles at Lawrence Livermore Laboratory (also located in California). A visible result of that cooperation was an attempt to explain the ozone hole (together with Solomon *et al.* 1986). Crutzen, who had established his success in Germany, cooperated with Brühl at the Max Planck Institute in Mainz and with Arnold in Heidelberg.

As the models became more complex and cooperation between experimenters and modellers grew, research groups tended to grow bigger. This is indicated by the average number of authors per paper.[13] In 1985 it was 2.7; by 1995 more than four.[14] In 1985 more than 95 per cent of all publications in the field were produced by six authors or fewer. In 1994 about 95 per cent of all publications in the field were written by ten authors or fewer. However, in both periods the most frequent form of cooperation is two authors. Today an integration of modellers, experimenters and technical support people is called for. If groups do not have the critical mass, they are unlikely to be on the cutting edge of research. Several interviewees expressed such a concern about the German research institutions:

> The German malaise is institutional. You know, also in the US you have different groups, but they are talking much more to each other. In Europe somehow there is secrecy. . . . The Germans have nothing to show. They have measurements, some interesting results. There are also individuals who have a good reputation . . . but the competence is scattered.
>
> (UNAS 2)

> Three years ago [1991] the working group stratospheric chemistry was upgraded to an institute so that the group could also attend to modelling tasks, in order to model their own data. This included an increase in staff. . . . For a long time the modellers and experimenters did not communicate very much. They worked along on their own. Synergy effects have only gradually developed, after the government created the programme which basically meant that cooperation was enforced.
>
> (GEAS 35)

> You can't start taking measurements on a green field. *In situ* measurements are difficult. One reason is the constraints which are posed by the instruments. You need the know-how for these. The same holds for the modelling, there was virtually nothing existent in Germany. When I arrived, I brought

with me Crutzen's initial 1-D model, that was the only stratospheric model which existed at that time in Germany.

(GEAS 26)

We started catching up with the USA not before the 1980s. We did not have systematic financing, we only had those small university institutes which were not staffed properly. We did not participate in putting together the international research programmes. ... Our cooperation was mainly transatlantic, rather than within our own country.

(GEAS 1)

Little wonder, then, that in comparison with the United States, Germany has shown little innovation. With the exception of the groups around Crutzen and Ehhalt, there were no important contributions for a long time. Of these two, one group mainly limited itself to modelling, the other to field experiments. There were very few interdisciplinary research groups, and those that did exist looked very much towards the USA.

Europe lacks the technological infrastructure. The scientist has to look after his data support network, that it is built up at the place where the experiment takes place. The Americans had many technicians who were just waiting for the scientists to return, handing over to them a tape with the data. The technicians copied that onto a mainframe, did the formatting and plotting; then the scientists returned and talked about the science. We have another tradition which is not necessarily wrong, but less efficient. The scientific groups do everything by themselves, starting with the putting together of the instruments, the wrapping, data transmission and analysis and finishing with writing the papers. We are much more intensive in all steps, simply because the infrastructure is lacking. There are very few areas where we have enough technicians and support staff so that the scientists may concentrate upon science and planning. The groups are all below the critical threshold.

(GEAS 35)

Apart from different national research conditions one important aspect has to be mentioned. The emergence of interdisciplinarity lets the field grow and shrink at the same time. It is getting larger because researchers from more than one discipline get together, and it gets smaller since they represent only a small part of their original discipline. This means that they are likely to be decoupled and even estranged from the rest of their discipline. Scientists who work in interdisciplinary teams tend to seek contact with other groups that also collaborate across boundaries. Sometimes, this is likely to lead to sever the relationship with the original discipline. Interdisciplinary cooperation therefore does not abolish disciplinary boundaries but creates new ones (Fish 1994).

Internationalisation

This paradoxical process of simultaneous expansion and contraction also applies to the formation of international research teams. The field is growing through cooperation with researchers working on the same problem in other countries; however, the decoupling from the old discipline lets it shrink as it gains in autonomy. A process of estrangement is the likely result. As with the interdisciplinary orientation, internationalisation seems to be natural, given the global nature of the problem. And as with interdisciplinary orientation, this did not happen automatically. It took a conscious effort. A key figure in this process is Bob Watson who acted as chair of NASA's programme for ozone research. In the beginning of the 1980s he rightly saw that CFC regulations would be hampered by the existence of many differing ozone assessments. At the time there were six different reports on the state of knowledge on ozone. This could only lead to confusion and, above all, it gave the opponents of regulations welcome arguments. These reports were commissioned by the European Community, NASA, NAS, UNEP, WMO, and the British government. 'At that stage industry and other people were looking rather at the differences than at the commonalities of the different studies. So I tried to work with the international science community toward a single international assessment' (Personal communication with Watson).

Watson successfully led the international scientific community in producing one single report. The first report was published in 1986 (WMO 1986) with several other reports following (WMO 1988, 1989, 1991, 1994). This reporting system provided a mechanism that allowed the bringing together of all the relevant scientists.[15] This was intended to overcome many controversial points and to bring them together in a common position. In 1992, Watson received the *National Academy of Sciences Award for Science Reviewing* (he had been nominated by Rowland). Eugene Garfield commented on this award in *Current Contents*:

> Watson . . . has been described as a 'national asset' . . . [He] has supplied the evidence to a sceptical world that proves there is such a thing as ozone depletion in the upper stratosphere, particularly over Antarctica, and that it threatens humankind's well-being.
>
> (Garfield 1992: 5)

The reason for giving the award to Watson was that his publications and the reports inspired by him were the basis for decisions made by industry and governments to control CFCs. Garfield stresses the fact that, although Watson's work was cited by other scientists, its main influence was on the political decision-making process.

In spite of internationalisation there are different national research programmes as well as national pride about national contributions. The existing national differences in the development of knowledge about ozone and the atmosphere are mainly due to institutional factors. Around 1970, there was a common knowledge basis in stratospheric chemistry, dating back to the 1930s.

When CFCs and other substances moved to centre stage, this knowledge became revised – but differently so in different countries. The international scientific community in ozone research is grouped around a core located in the United States, all in close contact with each other.

> They all know what the others are doing. That is the way we make the assessments. We count on these key people and after that they have groups around them and know others. . . . The Americans didn't involve too many Europeans . . . we have all the time tensions because of this.
>
> (UNAS 2)

These scientists also form the core of experts who contribute to the international WMO/UNEP reports. They form an 'invisible college' (Crane 1972; Price 1963).[16] In quantitative terms, the dominance of American scientists is quite evident (see Table 2.1).

This is in line with the worldwide structure of science. Several studies have shown that the United States is the world's leading science nation. Based on a co-citations analysis, Winterhager *et al.* (1988) found that the USA is by far the leading science nation: nearly half of all publications in scientific journals originate from the USA. Based on network analysis, Schubert and Braun (1990) analysed international scientific cooperation for the period 1981 to 1985. Their results show that the United States occupies a central position within an international network, although the degree of cooperation between American scientists and scientists from other nations is low. This indicates that the USA has a leading position in research and can afford to refrain from cooperation with scientists from other nations. In the case of ozone research, however, this *splendid isolation* was broken for political reasons. Constructing an international

Table 2.1 Nationality of authors and referees of UNEP/WMO reports

	1985	*1988*	*1989*	*1991*	*1994*
USA	99	104	81	128	143
UK	12	6	11	14	28
Germany	10	3	12	7	29
France	6	2	8	4	14
Belgium	4	2	3	1	4
Norway	0	1	2	3	8
Canada	2	2	2	2	2
Japan	0	1	5	4	6
Russia	0	0	4	6	5
Australia	1	2	2	1	5
Italy	2	1	0	2	3
Sweden	0	1	0	1	1

Source: WMO, own calculation.

Note: Sample of important countries.

institutional consensus requires the involvement of scientists from many different countries.

> Take these WMO reports. Who wrote them? There is no institutionalised democratic system to hire the scientists. Basically the way it works, WMO asks the US government to write a report. So they ask W. and A. They make an outline, like tropospheric chemistry: 'Whom should we ask?', and so on. Then they have the list and see that there are only Americans, that doesn't work. So they replace a few people with people from Europe. Now it looks international. But they are all part of the same planet, you know what I mean. So they say, we need to be more international, so they add more people, which do not have to do any work, but which are there on paper. 'Let's put a Chinese, three Japanese and an Indian. . . . ' And then they call a big meeting where all those people come from all over the world, they pay their trip and then: Who does the work? The lead authors who have been involved in the business.
>
> (USAS 41)

The goal of involving scientists from many other countries has only partially been achieved. Countries listed in Table 2.1 have increased their proportion, in some cases by a factor of three or more, but in absolute numbers the American scientists have gained even more weight. The proportion of non-American scientists is between 17 and 38 per cent. In 1994 it reached an all-time high of 42 per cent.

However, there is a significant shift in the relation between the lead authors for chapters in the WMO reports. In the 1988 report there were 14 lead authors, 10 of whom were of US origin, one each from Norway, Australia, England and Germany. In the 1994 report, out of 16 lead authors only six came from the USA, five from Great Britain, two from Germany and one each from New Zealand, Norway and Venezuela. Incidentally, a leading American atmospheric scientist thinks that in 1988, the situation was very much alive; by 1994, the major controversies were all over. Since the outcome was clear it wasn't so important to have the very best persons in the lead author positions – 'very good' could do almost as well. 'In basketball it's known as "garbage time": the decision as to who will win is already in place, and the substitutes come in, very good players, but not quite as good as the first team' (USAS 16).

The institutionalisation of the WMO reports was an important means of moving the international political process in the direction of CFC controls. Following their existence they tried to establish a consensus in the international scientific community. These reports were used as the main scientific legitimation for ozone regulations. As we shall see, they were not the main cause for bringing about these regulations.

Scientists as social actors

In decisions under uncertainty many people, including politicians, tend to over-look the fact that the scientific community itself is deeply divided. Instead, they seem to assume that the issue can be reduced to a purely technical problem with scientists giving ultimate and compelling advice. In the case of BSE, this led to one of the most severe political crises in the UK. The Conservative government acted timidly in the late 1980s. One of the reasons was that the scientists who sat on the government's advisory committee had their own perceptions of what was politically digestible (Wynne 1996b). The advisory committee did not advocate the need for a ban on the use of cattle offal in human food until November 1989. The chief scientific adviser recalled that this 'was a no-goer. The Ministry of Agriculture, Fisheries and Food already thought our proposals were pretty revolutionary' (cit. in MacNaghten and Urry 1998: 259). This clearly indicates that scientists' world-views, their perceptions of social and political affairs, are as important to the understanding of such controversies as are the claims of their knowledge.

World-views of scientists

On the basis of my interview material I classified the atmospheric scientists' viewpoints ($N = 27$) on environmental matters (ethical values, assumptions about important causal relations, political options). The sample comprises about 10 per cent of the international community, which can be estimated to consist of 200 to 300. However, this is not a random sample. Central scientists, who are especially vocal, visible and enjoy a high reputation, are over-represented.[17] This was intended, as I hoped to delineate the influence of scientific expertise on the policy process where I assumed that the visible and those with a high reputation would be influential. I distinguish three groups of scientists with regard to policy issues: advocates, sceptics and undecided. Advocates and sceptics were additionally classified as either vocal or quiet. In total, this yields five main profiles: vocal and quiet advocates, vocal and quiet sceptics and undecided (bystanders – by definition, they are not vocal). Vocal advocates are those who spoke up in public and gave policy advice; quiet advocates are those who engaged in policy advisory activities, but were much less in the public sphere. Similarly, vocal sceptics were active both in the public and in the policy process. Most importantly, they did not think that the link between CFCs and stratospheric ozone depletion was proven. They spoke out against quick or strict CFC controls. Quiet sceptics did not profess this position publicly. The undecided were not visible, they waited for data and proof. If one adds the time dimension, we note that there are *early* (i.e. around 1975) quiet and vocal advocates and sceptics. However there are neither sceptics nor undecided around 1986 (see Tables 2.2 and 2.3).[18]

Apart from these data, research in the *Science Citation Index (SCI)* of the Institute for Scientific Information showed the relative reputation of these scientists over the last 20 years.[19] This information was collected to determine whether there was a connection between scientific reputation and political influence.

Table 2.2 Profiles of scientists and their distribution, 1975

Advocacy	High		Low	
Pro	Visible advocates	3	Quiet advocates	6
Contra	Visible sceptics	2	Quiet sceptics	0
Neither	—		Undecided	16

Source: Own survey.

Note: [N = 27].

Table 2.3 Profiles of scientists and their distribution, 1986

Advocacy	High		Low	
Pro	Visible advocates	7	Quiet advocates	17
Contra	Visible sceptics	0	Quiet sceptics	3
Neither	—		Undecided	0

Source: Own survey.

Note: [N = 27].

During the 1970s, two stable groups of scientists acting as either advocates or sceptics can be identified.[20] These are three vocal advocates and two vocal sceptics. The sceptics change their position in the second half of the 1980s when both thought that CFC controls were justified. Interestingly, none of the early advocates becomes a sceptic but some quiet advocates become vocal advocates. Different groups influence the policy process with different strategies at different points in time. In the 1970s, this was the group of vocal advocates who enjoyed broader support after 1985, especially the two quiet advocates who organised the international process of scientific cooperation and policy advice.

The disciplinary origin of scientists is as follows: 52 per cent of all interviewed scientists were chemists, about a quarter dynamicists and a quarter from other disciplines. Among the leading eight scientists (measured by the frequency with which their work is quoted in the literature) there is a different picture: here we see only one dynamicist, 62.5 per cent chemists and a quarter from other disciplines. The professional specialisation is more balanced. In the whole sample I found 56 per cent modellers (44 per cent experimenters), with the leading eight split equally between them.

Responding to the questions about environmental values and cause–effect relationships, most scientists thought that nature in general was quite robust against anthropogenic changes, but not with respect to trace gases like CFCs. In the 1970s, very few believed that an environmental hazard caused by CFCs was possible. Accordingly, most scientists were undecided about policy options. This changed after the discovery of the ozone hole. From about 1986 nearly all researchers in the field joined the advocates (see Figure 2.3).

In 1975, fewer than 40 per cent of the interviewed scientists thought that quick CFC controls were necessary, 20 to 30 per cent were engaged in public debates and gave policy advice. In 1987, nearly all scientists were in favour of CFC

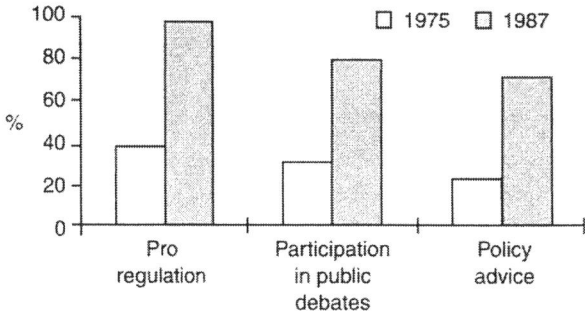

Figure 2.3 Scientific advocacy, 1975 and 1987

Source: Own survey.

controls, with more than 70 per cent participating in public debates and policy advice.

Ten scientists said they were in favour of CFC controls even before 1985; 14 were not active in the field or had no opinion. Only two were explicitly against regulations. While the position of these two is well documented in the literature, it is much more difficult to check the professed beliefs of the other group. It is possible that they have vetted their own position with the benefit of hindsight. A group of the scientists said that they were wavering when considering policy recommendations. They explained their indecision with references to the scientific uncertainty at the time.

From 1987, scientific uncertainty is no longer used as an argument against regulations. Dynamicists could have played this card, like the early sceptics in the 1970s – but they did not do so. (The same could be said about proponents of the solar cycle theory; I neglected them throughout the study as they only had a marginal position in the controversy.) Even the dynamicist who held on to his position longer than others, professed his adherence to the precautionary principle:

> I always thought that in the face of uncertainty one could take a prudent course of action just as a form of insurance, just like you are buying a fire insurance, you are not predicting that you'll have a fire, but if there is a possibility for fire you can take out an insurance.
>
> (USAS 38)

These findings indicate that cultural theory is not very convincing when applied to the case (see Chapter 1). In particular, it does not help to understand how scientists switched from a view of 'nature is robust' to 'nature is fragile'. There was no corresponding change in social position. These scientists did not leave their hierarchical academic institutions and join egalitarian environmentalist groups. Likewise, it does not explain why scientists had differing views. Only

Lovelock seems to fit nicely into the category of 'individualist', perceiving nature as robust.

When do scientists become activated?

In the 1970s, the undecided scientists were the largest group. If one applies Granovetter's threshold model (1978), the interesting question to be asked is when do scientists become activated? Granovetter sees the utility of his model above all in the analysis of social situations in which many actors act contingently upon the actions of others and where no institutional precedent exists to guide them. Thus 'bandwagon' effects are taking place. Reputation and 'seals of approval' are crucial here.

In the ozone case we observe the following distribution of threshold levels among the atmospheric scientists. A small group takes on an early advocate role and tries to get support from colleagues. However, their threshold level is higher. Before they become activated, they want to see evidence more compelling than the Molina–Rowland hypothesis.

> A lot of people were waiting to see how it comes out. For most scientists in most areas, they don't want to speak up unless they have really made an in-depth study of the area, so that they can speak on it authoritatively. The number of chemists/meteorologists who would know the chlorine chemistry and the meteorology was very, very small at that time. So, you would not expect very many people to speak up. They might say: that sounds interesting, even plausible, but if you believe it is a different question.
>
> (USAS 16)

> Most scientists preferred to stay quiet, for very good reasons. Physical scientists are trained to not make mistakes, to be very careful. And then there is an other kind of suspicion about being too public a personality: a scientist who is reported in newspapers all the time instead of writing journal articles and books is treated suspiciously by his colleagues.
>
> (USAS 5)

As I will show in greater detail, the activation of the undecided did not occur after the first report of the National Academy of Sciences in 1976 but after the discovery of the ozone hole. The NAS report, which advocated regulations, had a greater impact on politics than it had on other scientists.[21] After 1985, the higher threshold was passed by nearly everyone in the field. What was the trigger? My interview material strongly suggests that scientists came to realise that the system of the atmosphere was not behaving according to the theories, parameters and descriptions they were working with. These seemed utterly inadequate. Basically, the atmosphere behaved in an unforeseen manner. This was a huge shock that led many to speak out publicly.

You can't publish your paper and go home, and the all the rest of it has nothing to do with you. Everyone has a little line and once he is pushed over it, you suddenly realise: not only are you a scientist and having funded your work, but it is actually your job to get up and preach to people. And so it happened. . . . In a funny way, the ozone hole really has changed the whole environmental issue. The world we live in now is a different world than that I was born into.

(UKAS 44)

Once they realised that their received theories and models were all wrong, the scientists started to see a whole range of questions in a different light. It started with a radical doubt about their own atmospheric models. Even those researchers who were convinced of a significant ozone loss in the (distant) future were surprised by the dramatic and sudden events:

I do remember a meeting in '78, a number of the modeller groups got together. . . . We went around the table and asked what people thought the change in ozone would be in reality, not from what our models were saying, because we knew our models had their own problems. There were a few people who thought that there would be no real net change, even an increase. But most of us were saying the change would be big, 30 to 40 years looking into the future. And there have been much larger changes than we expected, because we did not have the understanding of chemistry and physics we needed to have.

(USAS 47)

The concept of nature changes accordingly: those who thought that nature was robust *vis-à-vis* human intervention changed their view either completely ('nature is not robust'), or else qualified it ('nature is robust but only within limits', 'nature is robust but not against industrially produced trace gases', 'nature is robust but not society which has to cope with the effects'):

I have the feeling that nature is amazingly robust on a global scale. There are all these feedback mechanisms that save you. But I also know that sometimes you are going over a bridge and you are getting into a positive feedback. When 'God' built the earth, he might not have thought about CFCs.

(USAS 41)

Even now I believe it [that nature is robust], but not for the ozone issue. You change the composition, the initial conditions, there is no way to be robust, to be able to recover alone.

(UNAS 2)

Before 1974 I thought it was quite robust and because of that, which was an implicit judgement, I questioned my first calculations when I realised that there could be a global effect from these industrial chemicals. My bias was that it was very unlikely given the large scale of the mass of the atmosphere.

And it's only by looking at the amplification factors that I changed my view. I realised that nature might be robust in certain aspects but not in all of them. Like any complex system it has some weak spots, the ozone layer is one of those. There might be other spots.

(USAS 13)

Around 1986 the environmental concern among the scientists increased, while the belief in the robustness of nature decreased. This is paralleled by a tendency towards advocating strict and quick CFC controls. Here the professional speciality seems to have exerted some influence. Chemists tended to rate the role of CFCs in ozone destruction rather high while dynamicists did not. Their professional beliefs made them sceptical about a causal role of CFCs in the emergence of the ozone hole; their professional bias made them spontaneously favour a dynamic explanation. Two of the interviewed dynamicists also told me that they did not consider themselves environmentally conscious. As a consequence, they did not speak up, warning against the dangers. But two chemists did exactly this, engaging in political and public contexts, although they told me that they were not environmentally conscious. This indicates that the professional background was more effective than the environmental/political belief system. However, those scientists soon turned into advocates, thus changing their positions about environmental and political questions as well. Only a few scientists said that they were environmentally conscious at the very beginning of the controversy. Even fewer said that this orientation had an influence on their work. The majority became politicised (if only willy-nilly) when they joined the field of ozone research. Finally, more and more scientists became interested in the political implications of their work. They participated in the discussion within science and society, contributing to the international WMO/UNEP reports, giving policy advice and making statements in public.

In sum, the threshold levels in the field of atmospheric research were distributed asymmetrically between two different groups: there was a very small group of scientists with a low threshold value and a large group with a high threshold value. Once the environmental conditions changed in a way so as to reach the high threshold level, a comparatively large group joined the avant-garde. This had an impact on the policy process. Scientists alarmed the public and politicians tried to get regulatory action under way as soon as possible. The consensus they had arrived at was rather normative-political, not scientific. They were agreed that one had to act even when there was no scientific explanation available. It was only in 1988 that scientists achieved a consensus about two key issues: the explanation of the Antarctic ozone hole and the observation of negative global ozone trends.

The 'backlash'

There is counter tendency against the dominant scientific view that gained some publicity in the 1990s. Its main points are as follows:

- Natural causes of ozone depletion are more important than anthropogenic causes. These are in particular chlorine emissions from volcanoes, the burning of biomass and salt from the oceans.
- CFCs are heavier than air and thus cannot rise to the stratosphere.
- The natural variation of the ozone layer is so strong that no significant trend can be seen in the historical record.
- The scientific community of atmospheric scientists has exaggerated the dangers in order to raise research funds (cf. Taubes 1993).

As I argued earlier, the enormous gain in knowledge in the field of the atmospheric sciences and the emergence of an interdisciplinary research community has led to a partial separation of this community from the mainstream of chemistry and meteorology. In addition, the institutional form of the policy network has helped to create the impression that certain scientific positions used to legitimise regulations are the result of wheeling and dealing, that science has 'constructed' facts. The EPA and NASA are particularly seen as conspirators. Well-known scientists cast doubts on the research results and the dominant paradigm in the field, among them the former president of the US National Academy of Sciences, Frederick Seitz:

> Frederick Seitz, from The George Marshall Institute, says 'The question should be emphasised that . . . Freon gas is much heavier than air.' [Which is supposed to show that CFCs do not rise into the stratosphere] He is the former president of the National Academy of Sciences. It sounds as though he should know something about it, but he doesn't.
>
> (USAS 16)

However, there is also a non-scientific reason for the backlash. It is striking that some backlash authors (the 'populists') attack those elements of the dominant paradigm that seem to be most firmly established: the fact that CFCs rise to the stratosphere and that their ozone depletion potential is much more important than natural sources. It is established by observations that CFCs do rise to the stratosphere and that they are quantitatively much more important for ozone depletion than natural sources. Although CFC molecules are approximately five times as heavy as air, they are mixed with other gases in the atmosphere, independent of their specific weight.[22] Since Lovelock's 1970 measurements of CFCs, atmospheric scientists knew that these substances would go up to the stratosphere. In 1975 this was confirmed by balloon and aircraft measurements. Since then scientists have carried out thousands of such measurements that have left no doubt about the fact that CFCs and their decomposition products are present in the stratosphere.

However, it seems that, for example, Singer and Elsaesser no longer question these basic physical elements of the current consensus. Singer did in 1989, but not during a survey performed by *Science* magazine's editorial board in spring 1993. Elsaesser thinks the problem is non-existent. Since the 1970s he has been of the opinion that 1 per cent more UV-B radiation increases the risk of contracting skin cancer by the same amount as if one were to move 25 km South (Elsaesser 1994: 44; cf. Dotto and Schiff 1978: 283). Therefore, political regulations are not imperative.

The second claim of the populist backlash rests on the assumption that volcanoes emit much more active chlorine into the stratosphere than CFCs. The argument runs as follows. In a publication in *Science* in 1980, a vulcanologist calculated that during its outbreak in 1976 Mount Augustine emitted approximately 175,000 tons of hydrochloric acid. The author estimated that during an outbreak of Bishop Tuff in California, 700,000 years ago, as much as 289 megatons of HCl could have been emitted into the stratosphere, which would amount to 570 times the amount of chlorine as the CFC world production in 1975. Ray and Guzzo (1990) take this number and apply it to the Mount Augustine outbreak of 1976; Limbaugh goes one step further and applies similar figures to the outbreak of Mount Pinatubo in 1991 (Taubes 1993: 1582). No matter whether this is due to negligence or to fraud (Limbaugh is host of a television show and not a scientist), the consensus among atmospheric scientists is that volcanoes do not emit nearly as much chlorine into the stratosphere as CFCs. Their emissions are less important since most of the HCl is washed out of the atmosphere by rain; the same applies for salt contained in the spray of oceans, but not for CFCs, which are not soluble in water (cf. Rowland 1993; Taubes 1993).

Obviously, the revisionist scientists and speakers are not as concerned about scientific discussion as they are about public policy. Hence, they direct their arguments not at the scientists but at the public. The backlash is not a scientific one but primarily a political phenomenon. It tries to exploit a sentiment in the public, which sees existing environmental regulations as exaggerated and too costly. This public sentiment can be found in the mass media and in public opinion, as numerous studies have shown.[23]

When this sentiment combines with scientific (or seemingly scientific) arguments, policy makers can face a problem of legitimation. It is difficult for lay people to scrutinise scientific arguments. But even supposing that this problem did not exist and they could understand all-important aspects, in many cases people simply do not want to believe that there is a problem. As a scientist told me, 'You just do not have the time to read everything that you should. Some of this backlash is not necessarily malicious. People want to believe that there is no problem' (USAS 5). Lay people are not only people outside science but also scientists outside the core group of scientists. They do not bother too much about the details of the issue. They firmly believe that in the past there has been an overreaction to environmental problems. Evidence for this was shown by an *ad hoc* survey carried out by the editors of *Science*. They asked scientists who were impressed by the book by Maduro and Schauerhammer and had signed a petition to revise the Montreal Protocol. The following statement by one scientist seems symptomatic: 'I'm one of those people who are opposed to getting scared about imaginary problems. I think the ozone hole and global heating are nonsense' (cit. in Taubes 1993: 1581).[24]

A brief glance at the book *The Holes in the Ozone Scare* by Maduro and Schauer-hammer (1992) discloses the specific character of the arguments presented. Apart from 'proof' that CFCs do not rise to the stratosphere, but natural sources emit chlorine into the stratosphere instead, it is claimed that the lifetime of CFCs (which has been established to be between 75 and 120 years, see chapter 3) is controversial; that there had been an ozone hole in former times; that the world-wide ban of CFCs would lead to the death of millions of people in the developing countries. The political rhetoric is bizarre: it talks about a 'green Gestapo',[25] a plot of neo-Malthusian scientists, irresponsible companies (Du Pont) and European aristocracy. The authors think that environmentalism is affected by paganism and Satanism; its mentors are identified in James Lovelock and Margaret Mead. The book ends with a description of large technological projects that are supposed to eradicate poverty.

3   Ozone controversies

Like everyday knowledge, scientific knowledge is not always free of passion and bias. The identification of important problems is not uncontested, and the same applies to proposed solutions. New areas of research in general are prone to scientific controversies, which is more likely the more applications or political implications the research has. Many observers believe that science will increasingly conform to contexts of application. If this is true, the lessons learned from the ozone controversy will inform our broader understanding of similar issues. Scientific careers are unlikely to evolve in separation from such controversies. This means, however, that scientists will increasingly have to build their careers on contested claims. And they will be much more motivated to defend their claims vigorously and not take an impartial stance, as the classical 'scientific ethos' would demand. To expect this would simply be to expect something super-human.

However, it would be wrong to conclude therefore that knowledge coming out of controversies would be unreliable. Quite the contrary: controversies are also beneficial. They attract the energy and resources necessary to settle competing knowledge claims. Without controversies we would hardly gain robust knowledge, since research findings would be much more scattered and incoherent.

In the ozone controversy, there have been several issues that were hotly debated. In the first decade, this was the long-term estimate of ozone depletion, in the second decade, there were disputes about measurements and causes of actual ozone losses. In this chapter, I shall first address normative issues of scientists' behaviour. Here sociology has developed some theoretical expectations that will be confronted with the empirical material. Second, I address scientists' prominence and reputation and their impact on the policy process. Third, I present important controversies that have developed around the ozone hole and show how they were settled. Fourth, I highlight the role of trust in scientific practice, and fifth, the relevance of rhetorical and visual means of representation.

Self-interest and norms in science

My analysis so far has shown that professional and career-specific reasons, as well as environmental viewpoints, influence research and related activities. This also applies to career planning. Today, as we observe self-interested students flocking into the 'hot fields' (like biotechnology) where huge financial rewards are to be expected (Stephan 1997), and norm-orientated students being attracted by the image of some sciences as 'environmental sciences', the same difference can be seen within the scientific communities, too. It can be generalised and related to two influential theorems of the social sciences. One might therefore ask: are scientists like other actors, utility maximising opportunists looking out for their self-interest with guile (Williamson 1985: 47; cf. also Boehmer-Christiansen 1994a and b), or are they different from other actors in that they follow the norms of universalism, communism, disinterestedness and organised scepticism (Merton 1973a)?

If one takes the view that scientists are primarily self-interested, one would assume that they strive to enhance their wealth. This can be done either by gaining lucrative positions within academia, or by gaining contracts outside academia and intellectual property rights (e.g. royalties, patents). The most important means of achieving this is by gaining visibility, either through scientific reputation (peer evaluation) or celebrity status. The usual way to advance is by being original or claiming priority to a discovery. Being original means doing something different from the vast majority of the community. Priority claims take this one step further. A priority claim is about a new piece of knowledge that is said to be stated for the first time. This is proven by publications in scientific journals. The reputation of a scientist is indicated by the frequency with which peers make reference to one's work. Science in general, but especially the natural sciences cite *new* knowledge.[1] Thus one can assume a close association between scientific reputation and the innovativeness of a scientist.

If one takes the view that scientists are norm-orientated, we have two possible scenarios. I introduced the first above where I showed that there is a connection between world-view (cognitive orientation) and practical-political outlook. The second is the view that postulates a specific ethic of scientific practice which is different from all other social spheres, as espoused by Merton. Merton's thesis contradicts the interest hypothesis. To be sure, Merton does not doubt that there are individual interests and that the breaking of norms occurs. But he asserts that the institutional incentives in science assure that these norms are upheld. Merton distinguishes between motivational and institutional aspects, but seems only concerned about the latter. This leads him to a problematical functionalist position.[2] We would get another picture if we started with the motivational structures and then analysed the macro effects they produce. This would have the advantage that one could show how norms influence behaviour via motivations.[3]

The distinction between norm-following and interest-following behaviour suggests the following analytical frame. Motivating norms can be identified in different views of nature. The early advocates clearly fall into the category of norm-following scientists. Here we find a strong correlation between their view of nature and their practical involvement. The three vocal advocates put it the following way:

> Our calculations [in 1974] made the situation very bad. The rate of growth of CFCs was enormous. There was a rate of 10–15 per cent per year, so we

had to assume a doubling time between 5 and 10 years. Then we had the natural delay time; the reservoir in the lower atmosphere had not been diminished for some decades, so it looked as if the ozone layer would be harmed. I was very concerned from the beginning and spoke out.

(USAS 5)

In my view it was not appropriate to release large amounts of anything into the environment without knowing what happens to it . . . I was in favour of regulations, although I clearly had to let a little bit of time go by to see how the scientific community would receive this first. Very soon thereafter it became clear to me, that we had to be advocates to this issue. We had to carry the voice for a regulation to happen.

(USAS 13)

If you have a bi-polar world in which you have two groups, the polluters and the environmentalists, then I am an environmentalist. But I don't think the world is bi-polar and I don't think of myself as being an environmentalist. I certainly don't think of myself as a polluter, but rather as a scientist interested in studying some of these problems.

(USAS 16)

Vocal sceptics are also professedly environmentally conscious. However, they do not advocate regulations. They act primarily to enhance their own interests, in other words, they primarily want to achieve career goals. It is no accident that they downplay the scientific consensus, while advocates tend to overplay it. One possible explanation would be that self-interested scientists try to maintain research money as long as possible since it is easier to legitimate additional research as long as not all important questions have been solved. Advocates, on the other hand, place a premium on the practical-political goals, thereby risking a dip in their careers.

Being original

Randall Collins terms the problem for scientists pursuing a career in the following way (he uses the term *intellectual* to denote scientists – in my view not a very happy choice of terminology):

The strategic problem of any intellectual is to be maximally original while yet maximally relevant to what the community is prepared to hear. . . . The basic problem of the intellectual career is recognition: how, then, does one make oneself visible when the sheer number of competitors increases by a factor of four or five?

(Collins 1986: 1337, 1339f.)

How is it possible for scientists to keep in contact with the discussion of the discipline and yet at the same time break free from it and attract attention? (Kuhn

[1977] saw in this the 'essential tension' of scientific work). The answer is by being original, and making important findings. Since this is part of a business in which everyone else is involved, the ensuing competition raises the question of *fair play*. To illustrate this point, I present the case of an atmospheric scientist who has been notorious for his drive towards fame. However, ironically, a strategy of being original nearly at any cost harbours the danger of being disavowed because of the high visibility (Zuckerman 1988).

Michael McElroy, a Harvard professor of great rhetorical talent, was among the first to research the stratosphere, especially the influence of chlorine and bromine. He was characterised as 'one of the most flamboyant personalities associated with the ozone controversy ... [He] is extremely competitive and favours a strongly confrontational *modus operandi* in scientific exchanges' (Dotto and Schiff 1978: 128).

Like Stolarski, he gave a talk during the above mentioned congress in Kyoto in 1973. For the purpose of publication, he modified the paper in such a funda-mental way that the main part dealt with the effects of chlorine (brought up by Stolarski at the conference) and not with NO_x which had taken centre stage during his talk. His close collaborator Wofsy had written a negative referee's report of Cicerone and Stolarski's paper during the review process for *Science*. On 26 September 1974, Walter Sullivan reported in the *New York Times* on McElroy and Wofsy's model calculations. This article scooped Cicerone and Stolarski's article in *Science*, which appeared the day after (Dotto and Schiff 1978: 23). This suggested an inversion of priority. It seemed as if McElroy and Wofsy had made the important discoveries before Cicerone and Stolarski.

Apart from Cicerone and Stolarski, Rowland and Molina were McElroy's main competitors. One is tempted to compare this competition to a Grand Prix race in which McElroy tried to manipulate the rules in his favour. It began before the real start when he tried to have others (Stolarski and Cicerone) excluded from partici-pating. After this attempt had failed, he tried to ensure coverage in the national press as *the* authority on ozone. In situations where the race is clearly dominated by others, he wants maximum attention by means of all sorts of performances. At times, it seems as if he lets himself be sidetracked before getting back into the race.

And he was quite successful in this game, although he did not get the ultimate recognition. For example, he drew the attention to bromine which is, like chlor-ine, an ozone depleting substance. He claimed that bromine was much more effective in destroying ozone, so effective that it could be used as a potential weapon (*New York Times*, 28 February 1975). The *National Enquirer* carried the headline: 'Harvard Professor Warns of ... the Doomsday Weapon ... It's Worse Than the Most Devastating Nuclear Explosion – and Available to All' (cit. in Dotto and Schiff 1978: 188). He suggested that bromine, not chlorine, is the substance that needs the attention.

More than a decade later, still in the scientific competition, he made use of his 'weapon' again, trying to explain the Antarctic ozone hole. After a consensus emerged in the community of atmospheric scientists that sun cycle theory and dynamic theory could not explain it, there remained several problems with the chemical explanation. Models based on it were initially not able to mimic the enormous ozone drop that occurs in a period of about three weeks. In this situation McElroy again advanced bromine, which in fact is a more effective catalyst than chlorine but exists in much lower concentrations in the atmosphere. Models of all researchers were forced to their limits (and beyond) in order to imitate these huge ozone losses. For nearly two years no one succeeded in doing so. Here McElroy, quite understandably, bet on bromine (today its relative significance is estimated at about 25 per cent). His work on the Antarctic ozone hole is widely quoted; he even achieved the highest increase in relative citation frequency from 1985 to 1990 – but he failed to be among the Nobel Prize winners in 1995.

Priority conflicts

Competition sparks off between individual researchers but also between institutions. Not only NASA claims to be number one. This is all within the range of the normal and expected. It becomes interesting as soon as the 'normal' is transcended. Some priority conflicts have been carried out using unethical means. This is well known in the scientific community of atmospheric research, but was not made into a major problem.[4] Scientists fight for the reward that goes to the first. Several priority claims have been made with respect to polar ozone chemistry. As an American scientist pointed out to me, 'Various people . . . wrote review papers, they decided to rewrite history in their favour or in whoever favour they wanted around 1988–90' (USAS 15). Another confirms this:

> Well, there were personality conflicts that had to do with the fact that there was a lot of publicity and some people reacted in understandable ways, made something more dramatic, that's all in the way ordinary people behave. A lot of that has to do with the question of who gets credit for what?
>
> (USAS 22)

Some people used 'dirty' methods. Again, McElroy serves as an example:

> He is an extremely brilliant guy and a very good speaker and he is able to take information from a lot of different people and blend it together and make it appear that he had solved the whole problem himself. I like him, he is a very good scientist, I had never directly to compete with him, you see. Some of the chemists think that he is sometimes unethical towards other people's work.
>
> (USAS 27)

> T. once said to me I should make a T-shirt that says: 'McElroy stole my ideas' – I could sell a million of them. It is just a joke, it is not something people take very seriously.
>
> (USAS 36)

He made people just mad. S.'s contribution was important, no one can take that away from her, but it didn't go far enough. But there is another part that belongs to T., and McElroy tried to steal this, I know this because I was the editor of a special issue. He called me up and yelled at me, you know, he the great McElroy. I just a dumb editor.

(USAS 17)

After the discovery of the ozone hole and the rapid political progress on the field of international regulations, it was not unreasonable to expect the Nobel Prize for outstanding achievements. This drove some researchers to perform special efforts in their research and publication strategies.

The publication dilemma

Merton's norm of 'communism' means that each scientist shares his or her results with the scientific community. This norm is likely to be violated if one fears that self-contradictory results diminish one's reputation, if one fears being 'scooped' by fellow scientists, if one can profit financially from the results of the research (Blumenthal *et al.* 1997; Press and Washburn 2000), or if one works for a client who formally forbids the publication of results. Likewise, if one anticipates large political effects one might be inclined not to publish if one wants to stay out of the fray. In what follows, I shall only touch upon the first two points.

Advocates and sceptics alike may obtain research results that do not fit their predictions very well. Sometimes new results make future ozone losses more dramatic, sometimes less. Depending on which side of a controversy the scientist stands, this is good or bad news. If one applies the distinction between norm-guided and interest-guided behaviour to the publication activity of the scientists, the Mertonian norm of communism would require that new scientific results are always published. This is also in the self-interest of scientists since this increases their reputation. This is what Merton suggests with his remark upon those 'happy circumstances in which moral obligation and self-interest coincide and fuse' (Merton 1973b: 399).

If we look at scientists who are not only engaged in a scientific controversy, but also in public–political controversy, this complicates the picture. New results may indeed contradict previous estimates of, for example, long-term ozone depletion. This is problematic for someone who had created a specific image, claiming, for example, that the problem was extremely serious. It is not unrealistic that she loses her credibility by publishing this new self-defeating result. The decision is a dilemma since one cannot know if the same result will be found (and published) by others. In case the own finding would not be confirmed by further research, one would have made much ado about nothing and done a lot of damage to oneself. [5]

For scientists in this dilemma a publication is only rewarding if others are likely to make the same finding. The least preferred options are either not to publish while others make the same finding and publish it; or to publish while the others do not make the finding. This calculus deviates from the standard model of proper scientific behaviour since it makes the decision to publish contingent upon the likely publication plans of others. The normal path on which one publishes whatever finding one makes is abandoned. An important variant of this dilemma comes in when publicity plays a role and scientists dramatise their results. The incentives to do so are clear: if one has alarming results, the research speciality has a good chance to get research money. In the end, an ambiguous incentive structure arises.

The peer review system sometimes leads scientists into the dilemma that they would like to publish research results in scientific journals but cannot do so for fear of being scooped during the process (see examples in Broad and Wade 1982; Chubin and Hackett 1990; LaFollette 1992; Mazur 1989; Shepherd 1995). This can also happen during 'normal' correspondence with colleagues. One scientist was about to make a big discovery (which proved to be fundamental) and eventually published it in a prestigious journal. He told me: 'I was lucky that other people with whom I was corresponding did not quite understand what I did, otherwise they might have published it before me' (GEAS 25).[6] These examples all indicate that there are powerful institutional barriers that counteract Merton's norm of communism.

Reputation and celebrity

Do highly reputed scientists have a special influence on the policy process? There are indications that this might be the case. In the UK, the expertise of the Royal Society is usually taken by government to represent true scientific knowledge on which public policy can be based. Likewise, the US National Academy of Sciences was seen by the Federal government as the final arbiter on the Molina–Rowland hypothesis. However, there are some important sociological distinctions that caution against a generalisation of these examples.

First, reputation is important for the inner scientific discourse where it serves to channel attention, thereby selecting scientific results and their reception by others (Merton 1973c). By contrast, people who play a visible role in public discourse are prominent, attaining the status of celebrities (Goodell 1977; B. Peters 1996; H.P. Peters 1994; Weingart and Pansegrau 1999). Scientific reputation can be measured by the relative frequency of citations, celebrity by the relative frequency of public appearances. Second, there must be a readiness on the part of the political system to receive information on topics that are defined as urgent by the scientists. Not every important issue finds a place on the political agenda. Chances increase if the agenda-setting process evolves with the participation of a large part of the public, especially the mass media. According to Mazur (1981), the mass media do not have the power to determine *what* people think, but what people think *about* – and this is decisive in order that an issue gains attention in the political process. If we define celebrity as the 'generalised ability of an actor to cause public attention' (Gerhards and Neidhardt 1990: 36), this suggests that prominent scientists possess more political clout than scientists who enjoy a high reputation.

In what follows I present findings from the present sample of atmospheric scientists. I calculated average relative citation frequencies for individual scientists based on their academic publications, as recorded by the *Science Citation Index* (see Figure 3.1). For the first period I took those five scientists who were most visible during the 1970s in the American public. Three of them were vociferous advocates, the other two were vociferous sceptics. The sceptics' reputation during that period is about double the reputation of the advocates. However, this was not reflected in the political decisions taken.

After 1985 a conspicuous correlation between the relative reputation and political influence occurs (see Figure 3.2.). At that time the chemical explanation begins to gain more and more weight and the early advocates get wide recognition. This affects their scientific reputation (see Bucchi [1998] for examples from other fields). However, as Figure 3.1 also reveals, the sceptics of the first decade had a higher reputation than the advocates. It is only after 1986 that this relation is reversed. For this reason, one has to exclude the hypothesis that scientists with a high reputation have *in general* a special influence on the political process.

In Figure 3.2 we see an increase in reputation by the advocates that in fact is reflected by the political decisions. A knock-on effect of scientific reputation onto the policy process would be supported only if one were to look exclusively at the data after 1986. If one looks at both periods, this conclusion does not hold. Rather, it seems as if the drastic shift between the two camps had a signalling function. The two vociferous sceptics from the 1970s went mute around 1985; one of them had withdrawn from the controversy while the other joined the advocates.

There is an important difference in the behaviour of the sceptics in both periods. In the first period the sceptics are vociferous; in the second period they no longer exist. To be sure, there are scientists who deny any correlation between CFC and ozone depletion, but they do not have a scientific reputation (they are

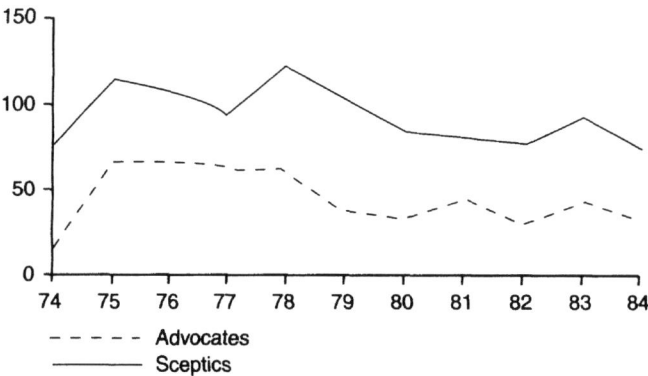

Figure 3.1 Relative citation frequency of advocates and sceptics, mean values 1974 to 1984, N = 5

Source: SCI, own calculations.

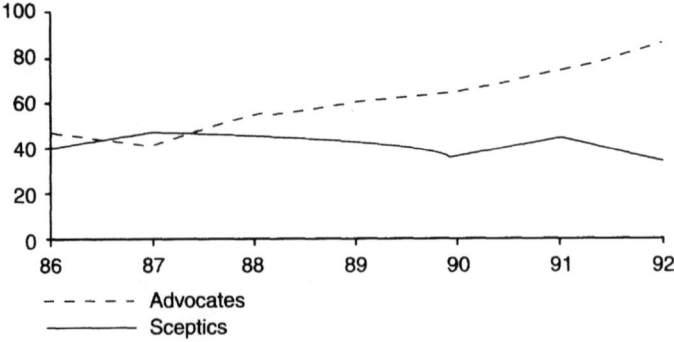

Figure 3.2 Relative citation frequency of advocates and sceptics, mean values 1986–
          1993, N = 12.

Source: SCI, own calculations.

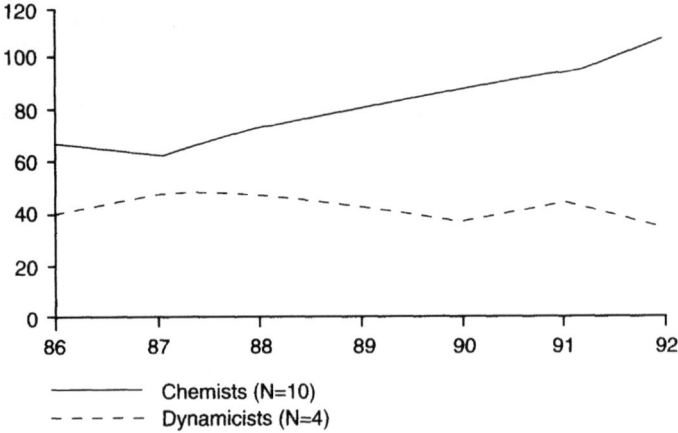

Figure 3.3 Relative citation frequency of chemists and dynamicists, mean values
          1985–1992.

Source: SCI, own calculations.

outside the field of active ozone research; see Chapter 2, 'The backlash'). After the
discovery of the ozone hole, there are scientists who remain sceptical with respect
to the chemical explanation. But they do not deny in public the necessity to curb
CFC emissions. The sceptics (N = 3) in Fig. 3.2 come from this group.[7]

Figure 3.3 shows the citation frequency of the exponents[8] of the ozone hole
controversy. The controversy lasted for only a relatively short time (1986–88).
After 1987 the chemists enjoy an advantage which remains stable over time. Their
growth in reputation is nearly linear, while the dynamicists stagnate. Since my
data basis for this calculation are authors rather than articles, this indicates that
the career prospects of the chemists have increased.

It is remarkable how strongly the field is characterised by leading scientists.

Their relative share in the sample (N = 27) is 50 to 70 per cent, whereas the expected value would be 30 per cent. Early advocates and sceptics (N = 6) have a share between 30 and 60 per cent (the expected value is 22 per cent, see Figure 3.4).

Involvement and objectivity of scientists

A problem arises for the scientific advocates: they are accused of betraying the ideals of science. Their public role entails 'popularising' scientific findings (Hilgartner 1990), taking sides in a social and political controversy, even, at times, making policy recommendations. However, none of the scientists active in this field of research could avoid asking (or being asked) questions such as: Who has the burden of proof? What is reasonable evidence of damage? Who should make judgements on these issues? How should one weigh 'worst case' scenarios? What weight should be given to social or economic benefits when considering regulation? (Brooks 1982). Those were also the questions that had to be answered when political options were formulated – scientists and politicians alike had to find responses.

The scientists who were involved in the CFC controversy 'found themselves unable to avoid making explicit or implicit judgements about almost every one of these essentially non-scientific value questions, no matter how much they tried to "stick to the facts"' (Brooks 1982: 206). Rowland gave priority to ecological concerns when asked to rank them with economic interests: 'I think that the economic dislocation need to be given minimal weight compared to the maximum weight to the possible harm to the environment.'[9]

Opponents of regulation aired the view that the burden of the proof was with the regulating agency. A speaker of Du Pont said: 'As a prerequisite for

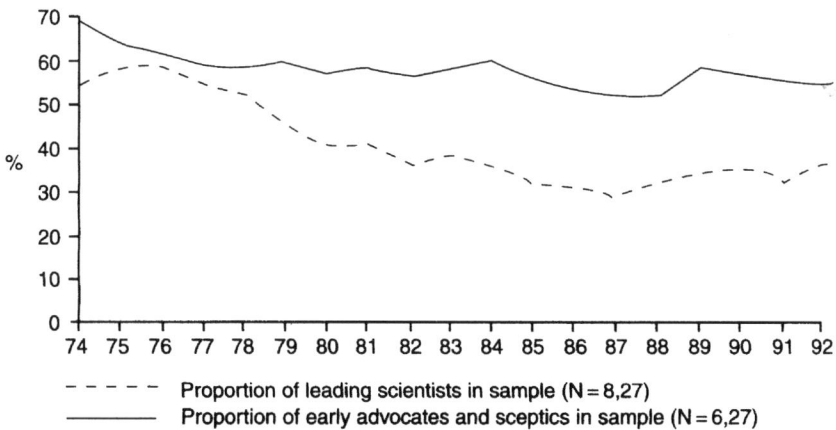

Figure 3.4 Citation frequency of leading scientists as a proportion of the sample, 1974 to 1992

Source: SCI, own calculations.

regulations, the promulgating agency should be required to affirmatively find a probable hazard, based on accepted scientific data.' He doubted the validity of the ozone 'hypothesis' and demanded facts, not 'theory'.[10]

There was little space for scientists who stood between the two camps. In fact, they were nearly forced to take sides. An early example can be found in the Congress Hearings of 1975 when senator Bumpers tried to push those scientists who tried to remain neutral, like James Anderson. Eventually, Anderson bowed to the pressure and came out in favour of preventive action, pointing to the 'very delicate ozone photochemistry'.[11] Another remained firm and gave elusive answers. He said, 'my advice to you on that issue isn't worth any more than the advice of any layman on the subject.'[12] However, in the end, he made a statement about the urgency to act ('Waiting a couple of years is not wrong') and one can be sure that his statement carried a bigger weight within the subcommittee of Congress than that of an ordinary citizen. Whether they want it or not, scientists have a special role in this process of providing legitimation for government action (Mukerji 1989).

Politicians who want to make use of scientific expertise have the problem that the body of expert knowledge is ambiguous. This problem is especially felt in the environmental policy arena. Politics, facing a value judgement, tries to pass the problem on to science, which is thought to provide a rational and reliable course of action, based on its 'objective' methods (cf. Brooks and Cooper 1987). However, this hope is disappointed most of the time; what we get instead is a controversy among experts. Within that controversy, value judgements play an important role, too, no matter how subtle. One could argue that the problem lies with the politicisation of the process of knowledge production and that 'pure science' could fulfil the high hopes if it were not polluted by political influence. One scientist who is not publicly visible in the controversy came up with this view:

> 'My expert says this, your expert says that', then you have two experts, this gets closer to the backlash. In a proper scientific discussion, usually we should be discussing things that can be resolved by observations. It's not a question of one expert versus another, rather one prediction versus another, but this does not happen when it is politicised.
>
> (USAS 22)

As I try to show, it is not that easy (cf. Weingart 1981: 218). The ozone controversy shows that although eventually a consensus emerged, this was *not in spite but because of politicisation*. The process was accompanied by public attention and an increase in resources for ozone research which was legitimised by the importance of the problem. As the world watched the ozone hole, scientists could not afford to stand on the sidelines. Those who did not participate in the public controversy played only a minor role in the scientific discussion. Only three of the interviewed scientists did not become involved in the public dispute. Nevertheless, vociferous scientific advocates must beware of appearing as 'mere' environmentalists. Purification is always at work, as the earlier quote in this chapter showed:

I am not a member of environmental groups, basically not then and not now
. . . I don't think of myself as being an environmentalist. I certainly don't
think of myself as a polluter, but rather as a scientist interested in studying
some of these problems.

(USAS 16)

Scientists view regulation authorities like the EPA and environmental groups as
political institutions from which they have to stay clear:

When science gets too close to policy then there is pressure put on the science
to short-cut and get to policy-relevant results. I think that's a big problem. In
the US we call that the EPA-effect. Because they are so oriented towards
regulation, if you get close to them scientifically, you'll stop to work for
science. They have their agenda.

(USAS 11)

No matter which side the scientists chose, all of them played the game of
purification, all stylised themselves as people who make their statements only on
the basis of the available scientific evidence. It is thus not surprising to see advocate scientists distancing themselves from environmental pressure groups like
NRDC[13] or Greenpeace.[14]

Scientists who took on an advocacy role show an ambivalent relation to activities of environmental pressure groups in the field. Some like their work and the
function they exert as pressure group, others see the consequences of emotional
over-dramatising: 'Greenpeace once came in space suits to a press conference
which was counterproductive' (USAS 8).

Very often I get application letters which say: I am very environmentally
conscious and therefore I want to work with you. I rarely take these people
since they are biased and you may fear that they falsify the data. . . . To my
knowledge, this has not happened yet. . . . They need to have really good
exam marks then I will employ them in spite of it. Once I had someone
working here who was a Greenpeace member. I gave him a project which had
nothing to do with the environment.

(GEAS 25)

Only those scientists who perform 'sound' scientific research are in a position to
convince colleagues and the public of their results. In order to do so, they must be
able to draw the distinction between the centre and periphery of scientific activity.
As Shils put it, the scientist

must remind himself constantly that his scientific knowledge is entitled to
agreement by others, non-scientists as well as scientists, only if it is scientifically true. . . . The scientists who engage in these penumbral activities have, for
the most part, been able to keep fresh in their minds the distinction between

the center of the scientific community in which scientific research, teaching and training are conducted, and the periphery of political, economic, and social activities with which the scientific community is in intense interaction.

(Shils 1987: 199, 198)

This quotation exemplifies how scientists approach the problem of public involvement. However, it does not address the difficult question of how scientists should behave in cases where they do not have established proof or evidence, i.e. where there is no certainty about the validity of scientific statements. Mario Molina, co-author of the Molina–Rowland hypothesis and an early advocate of regulations, thinks that the dominant view of scientists has to be challenged: 'They say "I'm just going to do clean, pure science. If it has some kind of application whatsoever, I'm going to step back and do something else." I think that's very wrong' (cit. in Roan 1989: 121). Their public restraint has several reasons:

Chemists have tended to feel stigmatised by all the adverse publicity that has surrounded their profession in recent years. Their reaction to environmental problems caused by chemicals . . . is frequently a defensive withdrawal from public involvement. Many of them are convinced that such problems are either non-existent or grossly exaggerated.

(Rowland, cit. in Cagin and Dray 1993: 306)

There are scientists who try to avoid the limelight by their very nature, and there are others who would like to be at the centre of attention, yet do not appear as advocates. The main reason not to do so is probably the fear of a loss of reputation which may occur in case some publicly made statement is not sustainable.

There is a debate about what moral obligations scientists have to take positions about what society has to do. I have very carefully stayed away from that, avoiding the non-scientific limelight. Sherry Rowland is a person whom I like very much and respect deeply, who crossed over the line and got away with it. But many people do not.

(USAS 11)

These scientists think it is their task to do 'sound science'. Among them, there is a group that I call the 'silent advocates'. Their strategy could be based on the rationale that the effect of one's own work is perceived to be larger if it is presented as neutrally and objectively as possible, even if one is guided by specific environmental thoughts:

I simply feel that if you're going to trust me to tell it like it is with the science numbers, then you'd better not question what my motivations are. My motivations are giving the best numbers, no matter what. . . . If I have any personal feelings, that has nothing to do with my credentials as a scientist. . . . I am one of the few who gets invitations to speak to the right-wing and left-wing groups, to speak about the science, because they trust me to discuss it without ideology. And I feel comfortable in that niche because it has given me a credibility that I would not have by advocating on one side or the other.

(USAS 11)

Upon questioning, this scientist did admit to be influenced by environmental values: 'If you were to accuse me of having a bias, it is a pro-environmental bias, yes. Although I try to keep that out of my statements and calculations. Environmentalist stewardship ethic is the way I would describe myself.'[15] Two other scientists who have been immensely influential within the policy process answered as follows:

My value to this process has been to speak as accurately as I can on the knowns and the unknowns of the science. My personal belief is separate from that in terms of being a citizen of a country.

(USAS 46)

I always kept myself in a situation that some people were not happy with and that is, I very purposely did not comment on whether there should or should not be international regulation.

(USAS 39)

The former did not want to comment on his private opinion: 'I would prefer not to answer. I would like to be able to be thought of as an objective scientist and I think it is that role that I play my best for the world.'

The reactions of both groups reflect their position within the policy process. The first quote is from a researcher whose main activity is within academic research, the second and third from scientists who are organising and coordinating international research and the presentation of this research to policy makers. They have to avoid being perceived as biased at any cost.

Scientists who did take an advocate role were thus ready to take a risk, as their more silent colleagues noted. Rowland, especially, has been reproached several times for not having the scientific data to back up his position.

His reputation might have suffered at a certain time because he was making lots of statements, he was probably perceived in general as what we call going out on a limb, i.e. making statements without having the certainty that they were correct. Here it is a matter of taste, I think. In the end, of course, he did the right thing because the theory was proven to be correct.

(USAS 13)

Ten years ago one could have said he was going out a bit on a limb. Of course he proved to be right. In the past I heard him make statements, that's true, but understandable when the chemical industry such as Du Pont was trying to avoid any regulation.

(USAS 27)

His involvement has been detrimental to his scientific reputation since he did not always distinguish between his scientific and political or personal opinions.

When I first read Rowland's and Molina's paper, I felt that this was a brilliant paper and they deserve all the credit they've got for it. And I was quite convinced there was a real effect. The only thing I felt uncertain about was the extent of it. That was my first reaction and then I heard Rowland giving lectures. He was like a missionary. He went around waving his hands, saying this is the end of the world.

(UKAS 42)

His colleagues also note that his strategy was very risky, albeit a rewarding one in the end:

Rowland, I think, made a very clear important decision, that a case had to be represented in an unequivocal way and I think in many ways he just sacrifices a lot of his scientific credentials to do that and I think this was a very import-ant decision that he made, not many people would do that. Maybe it paid out in the end, but at an extremely high risk.

(USAS 8)

From the beginning of the controversy, Rowland felt the effects of his involve-ment. It was little wonder that he was rarely invited to the chemical industry's scientific meetings. Universities were also reluctant to invite him. Exceptions were the award of the Tyler Prize for the environment (1983) and the membership award of the National Academy of Sciences (1978). His biggest reward came in 1995 when he received the Nobel Prize together with Molina and Crutzen.

Contradicting truths: the need for standardization

In the mid-1970s, American political decision-makers waited for the verdict of the highest scientific institution, the National Academy of Sciences. Its judgement was also received by policy makers abroad. In the USA it was seen as legitimising the ban on CFCs in aerosol spray-cans which was enacted in 1978. As soon as the problem entered the international realm, the NAS lost its status as quasi-Supreme Court, since expert panels in other countries partly reached other conclusions. For example, the Stratospheric Research Advisory Committee in the United King-dom found that the Molina–Rowland hypothesis was in doubt because of inac-curacies of computer models and a lack of understanding of chemical reactions in the stratosphere (UK DoE 1976; 1979). These reports were keenly taken up by CFC manufacturers and governments supporting them.

In the mid-1980s, the controversy continued in the institutional framework provided by UNEP and WMO. They had established a system of reports in which leading atmospheric scientists of several countries participated. Scientists are lead authors for chapters, authors of papers or referees. This process can be described as standardisation.

One of the most troubling problems scientists encounter when trying to establish knowledge claims is the standardisation of methods for measurement. The problem has been around from the oldest times when people disputed the size of an acre or the length of a yard. Perhaps the watch is the oldest scientific instrument that exemplifies this problem. In modern times, nearly every form of behaviour becomes precarious if it cannot orientate itself towards a unified standard of time (Giddens 1990). To maintain this standard, an international network of highly accurate atomic watches, lasers and satellites has been established under the supervision of the International Bureau of Time. 'As soon as you leave this trail, you start to be uncertain about what time it is, and the only way to regain certainty is to get in touch again with the metrological chains' (Latour 1987: 251). Scientific measurements, particularly on the micro or macro level which have to be comparable over long distances, have to be maintained by a network of negotiated standards. This process is impossible without standardisation, and standardisation is impossible without negotiations (O'Connell 1993; Porter 1995). No matter whether a census has to be done or the amount of emissions or global mean temperatures has to be established, or the validity of a laboratory experiment verified – each time there is an institution that is linked to a technical network or controls it. According to an estimate from 1980, the USA spends 6 per cent of its gross national product (GNP) on standardisation; this is three times the sum of research and development (Hunter 1980). Latour coined the term 'centres of calculation' for this infrastructure of science:

> The negotiation on the equivalence of non-equivalent situations is always what characterises the spread of science, and what explains, most of the time, why there are so many laboratories involved every time a difficult negotiation has to be settled.
>
> (Latour 1983: 155)

Assessing chemical reactions

In the ozone controversy of the 1970s the model predictions varied mainly because of the number and types of chemical reactions involved, their speed (rate constants), and the atmospheric concentrations of key substances. In the mid-1970s the NAS undertook the task of appraising the quality of input data that went into the model calculations (Dotto and Schiff 1978: 217). On the level of individual research groups, junior scientists did the work, as one atmospheric scientist explained:

> In the old days . . . you had to have at least one Postdoc. who was competent enough to review all the kinetics literature and start thinking up the best rate constants. And everybody calculated with different values. So someone had to help us out and do the review. . . . You don't have to re-invent the wheel. . . . The problems are getting messier so that's why standards are coming along.
>
> (USAS 15)

In 1979 a commission at the Jet Propulsion Laboratory (JPL) in Pasadena was established to assess the validity of rate constants that come from all over the world. Their data are binding for all research groups active in the field.

> JPL publishes [them] and Bill DeMore of JPL is the chairman. He is a very well-respected chemical kineticist. They do no measurements but evaluate all measurements from all over the world. . . . In addition to that, there is an international group called Co-Data. They lean very heavily on the JPL and there are no really serious discrepancies between them.
>
> (USAS 16)

> All these things are done practically in one place in the world and this is the JPL, funded by NASA. Everybody in the world uses their constants. They are investing millions of dollars for one reaction. When they have the numbers, they are good.
>
> (UNAS 2)

This process of standardisation and institutionalisation helped to block one possible source of controversy, namely the type of chemical reactions and differing rate constants. Another possible source of conflict was the discrepancy between satellite data and the data from the ground-based Dobson network regarding measurements of global ozone decline which became manifest after the discovery of the ozone hole and which I shall discuss in what follows.

The Ozone Trends Panel (OTP)

In 1985, interpreting data from the TOMS (Total Ozone Mapping Spectrometer) satellite, atmospheric scientist Don Heath of NASA found a significant decrease in global ozone. However, there were doubts about the validity of the database.[16] NASA checked the data and the instrument. NASA's Bob Watson commissioned Rowland to chair a working group to re-evaluate the data. Members of this working group not only included NASA scientists who had worked with the TOMS instrument, but also the experts for the worldwide Dobson network. The problem was that each measurement technique showed a different value.[17] As the working group commenced its work, the satellite experts tried to convince all others that their method was superior and hence their database was to be trusted. It was not before long that the Dobson specialists turned the tables. Walter

Komhyr, who was responsible for the American ground-based stations, explained to the other panelists in a detailed way how the Dobson instruments are calibrated on Mauna Loa (Hawaii), with its ideal atmospheric conditions for calibration:

> This has the best conditions to make a calibration, it is 20 degrees north and located on a hill. So if he [Walter Komhyr] could convince them [the TOMS scientists] of the Mauna Loa instrument, you could start talking about the other instruments, that was basically the psychology of it. So he went through these things, showed exactly how the calibration was made; we spent a morning or more on it, everyone tried to tell him why he was wrong, but no one could. And so eventually it was believed.
>
> (UKAS 43)

Komhyr succeeded in persuading the TOMS scientists that his calibration method was reliable and that it was the TOMS data that were flawed. As a consequence, the TOMS instrument on NASA satellite NIMBUS 7 had to be checked. It turned out that TOMS had recorded faulty measurements. The reason for this was an unrecognised technical problem with the diffuser plate which pointed at the sun:

> Before this problem was detected, the data was interpreted as increasing backscattered UV-radiation which meant ozone was going away. But it was simply the diffuser plate getting darker. Since the sun was used as source of calibration, it was not interpreted as darkening, but the earth as becoming brighter.
>
> (USAS 17)[18]

From this, the OTP concluded that it would have to rely exclusively upon the data from the ground-based stations. However, this created two new problems. First, the problem of reliability of the data of the Dobson network which itself suffered from problems of calibration and maintenance. Second, the problem that on the basis of its time series, there had never been a negative trend detected. As regards the first point, an experienced field observer made the following statement:

> This is the trouble of publishing all these data, you cannot go and pick them up and trust the data. It is completely impossible to evaluate if you can trust it or not. It is very sad.
>
> (UKAS 44)

His colleague agreed:

> While we looked at all the other stations, it was clear that data quality was a big issue. A three per cent change in calibration would cause a three per cent trend basically. . . . But the problem remained with all the other Dobson stations. One way of checking the calibration would be to bring them to

Mauna Loa, which is of course terribly expensive. Another way is to move that instrument to a location, but each time you move the instrument you risk messing up the calibration. So however you do it, there are risks. But you can do your own calibration on site, that's what Joe Farman did, it's just not as good as the one done in Mauna Loa. You then ask: How can you test the quality of the Dobson stations?

(UKAS 43)

To solve the calibration problem, the deviations of two neighbouring stations were compared over a long period of time and also compared to satellite data. If there was a sudden jump in one database that was not present in the other, this could be assumed to reflect a calibration problem. However, comparing calibrations creates new uncertainties, since the new calibration is an interdependent process and therefore can spread faulty values from one instrument or another – making it impossible to identify the source of error. If one were to avoid this, only one variable per time may be adjusted, which is very time-consuming. An important station (Halley Bay) that was not involved in the process of mutual calibration slowly worked towards a standard that made the scientists confident that it would guarantee reliable data.

We were lucky . . . we never were able to take part in these inter-comparisons. So we had to take jolly good care that our observations stood by themselves. So we actually did calibrate the instrument very carefully every year ourselves. Normally when you have these inter-comparisons, you mess up the instrument. So we were free of that problem.

(UKAS 44)

The second problem was that scientists working for Allied Chemicals had been analysing the long-term data in the mid-1970s but had not found any decrease in ozone concentrations. Even in 1985, they presented data that did not show any significant change (e.g. St. John *et al.* 1982). However, the data had not been updated and were only going until 1978.[19] 'We were in the summer of 1985 and they were quoting studies from 1979 with data stopping in 1978. So they had not updated their report' (USAS 16).

During an ozone workshop in the summer of 1985 in Les Diableret, the statisticians were not present to discuss their data. Rowland was angered by this, since all the other scientists had to listen to the criticism of their colleagues. Rowland himself had looked at the long-term data taken in Arosa which went back to 1926. His impression was that there was a slight downward trend. Neil Harris, a Ph.D. student working with him, analysed the Arosa data to find out if Rowland's impression was correct, and if so, why the statisticians did not see a downward trend:

So the statisticians gave exactly what the Chemical Manufacturers Association would like to hear. The CMA would not ask: Are you sure you have not screwed up and there is really ozone loss? So, I said to Neil Harris: I don't understand why St. John was getting an increase while my eyeball sees a decrease at Arosa.

(Personal communication with Rowland)

Harris soon discovered that there was no clear-cut difference in the data if one looked only at the annual means. His final conclusion in the OTP (Chapter 4) is just the opposite – that the loss is significant. However, the path to this conclusion started out more or less cursorily with the yearly data, then going quickly to the monthly data and finding the strong wintertime ozone loss at Arosa, then Bismarck and Caribou, then all 18 of the northern stations, all accompanied by uncertain summertime loss. Only finally did he go back to the yearly means after the great importance of the wintertime effect had already convinced the panellists that ozone loss was real, and that it occurred primarily in the winter.

Looking at the monthly means one could see a difference in the data taken before and after 1970. After 1970, CFCs were accumulating in the atmosphere, which by 1986 was a well-established fact in the community of atmospheric sciences. If the Molina–Rowland hypothesis was correct, an increase in CFC concentrations would lead to a decrease in stratospheric ozone. And indeed, by comparing seasonal data, Harris saw an ozone decline in the winter months after 1970.

What followed was a re-evaluation of the measurement of all Dobson stations. At its first meeting in December 1986, the Ozone Trends Panel set up ten working groups, among them a group to analyse trends in total ozone. Five scientists were involved in the main work: Rowland, Harris, Bloomfield, McFarland and Bojkov.[20] The statistician Bloomfield had done previous trend analyses. McFarland worked as a scientist for the Du Pont company and Bojkov was the specialist of the WMO for the ground-based network. He had the task to check the databases of all Dobson stations, to identify possible mistakes and to correct them.

Taking measurements with Dobson photospectrometers is not a trivial task. There are several possibilities for faulty readings, one of which is poor maintenance. Provided that all stations keep minutes about their measurements and maintenance, measurement errors can be identified and the data set can be corrected retrospectively (WMO 1993). As a British scientist told me,

You can't do anything in terms of improvement unless you have the original readings. There is a limit to what you can do, you can make consistency checks, you can look at the operating conditions, and you can say: This looks like a very silly value. But there is no way of actually finding out what went wrong unless you actually got the original sheets written down by the man at the time.

(UKAS 44)

It is hardly surprising that different scientists have different views about what are good and what are problematic data. The person responsible for the international network would like to do all the analyses by himself. However, the stations do not always provide him with the current data. For example, the people in charge of the British stations do not deliver their data within the set time limits for fear that they would not be given adequate treatment. They prefer to check their data themselves before passing them on to the official site. This mistrust is partially based on the fact that Bojkov had been unrestrained in his many corrections, which led to the allegation that he had manipulated data.

> We are much more successful than in the past, but we are clearly still not 100 per cent successful. Ninety percent of the changes Bojkov does, he can clearly justify. My worry is that he tries to tidy up the last remaining bit a little bit too much. It doesn't affect at all what the message from the network is. I am convinced that's right because he can justify so many of the changes. I am worried he is a bit too zealous for the last little bit.
>
> (UKAS 43)

> Bojkov went through a lot of these stations and made approximate corrections. This significantly improved the record. But it also made the record significantly more dependent on what one person did to it. . . . So you are very easily subject to the criticism of manipulation.
>
> (USAS 45)

But even without the corrections, 18 out of 19 stations showed a winter decrease while there was both increase and decrease during the summer. If one takes the 'corrected' values of the same 19 stations, they show a lesser decrease in ten cases and a stronger decrease in eight cases. The overall picture looks more homogeneous (WMO 1988: 241–4).

So, while the earlier trend analyses of the statisticians did not show any ozone decrease, the re-evaluation led to the opposite result. Why was there so little controversy between the statisticians (who worked for the chemical industry) and the atmospheric scientists (who included the most vociferous advocate)? I can see two reasons:

1   the early trend analyses did not show any ozone decrease because the most decisive changes occurred only after 1976, and the original data set stopped in 1978. Every additional year increased the chances to see the change, 'If you now do an annual average, you will see the trend. In 1987 to 1988 it was harder to see. It depends on how you handle various things. There were lots of detailed arguments' (USAS 45).

2   the method of trend analysis had been changed. According to the old method, there was a big standard deviation (compared to natural variation) during the winter months and a small one during the summer months. If one was to look for a change of ten Dobson units (DU), it was easier to see in the summer data than in the winter data. As a consequence, the statisticians weighted both groups of data differently: they gave more weight to the

summer data and less to the winter data. With this method they did not get a significant trend since the important winter data were diluted so much that they did not have an impact on the final result. The statisticians' tacit assumption was that ozone decrease would be monotonous. They thereby excluded the possibility of a seasonal ozone loss during the winter (USAS 16).

The statisticians who had done the early work were St. John (from Du Pont), Hill and Bishop (from Allied), Reinsel and Bloomfield (both academics, if funded by industry earlier on). Bloomfield was part of Rowland's group, intimately involved with the details of the new analysis. He changed his mind, agreeing that the new method was superior and was showing a real effect. Then, he talked with Reinsel and with the Allied people, and eventually persuaded them. St. John had dropped out about six years earlier. At the same time a change of heart of the American CFC producers occurred. It seems a legitimate speculation that industry would have tried to defend the 'old' method if there had not been a change in strategic orientation. However, this would have become very hard to do given that it 'lost' the key statisticians.

Decisive experiments

In 1986 and 1987 decisive experiments took place with the aim of exploring and explaining the Antarctic ozone hole. These experiments were extraordinary. They took place under exceptional climatic conditions, were planned like military action and on short notice, they demanded extreme motivation and readiness on the part of the researchers (open to heroic stylisation and dramatisation), and, last but not least, they eventually confirmed and falsified competing scientific hypotheses.[21]

The first Antarctic expedition, 1986

The first series of experiments took place from August to October 1986 during the National Ozone Expedition (NOZE) to McMurdo Station and was carried out by four teams with a total of 13 American researchers.[22] The experiments were coordinated by Robert Watson (NASA) and financed by NASA, NOAA and the National Science Foundation (NSF). Since NASA is not required to go through a time-consuming peer review process when preparing such experiments, Watson could pick the researchers and determine the type of experiments they should carry out. The chairman of the NSF was strongly in favour of the NOZE expedition. He felt that American high-tech science had to make up for missing the ozone hole which had not been discovered by NASA's 'multi-million fancy stuff' but by a thousand-dollar instrument based on technology from the 1920s (Cagin and Dray 1993: 294, 298).

The NOZE researchers presented their results on 20 October 1986 during a press conference that was broadcast live to Washington. The journalists were mainly interested in whether CFCs were causing the ozone hole. Susan Solomon, who acted as speaker of the expedition, felt it legitimate to answer the question in the affirmative. This provoked controversy as soon as the group arrived back in the USA (Roan 1989: 171–2). Exponents of two competing theories, the sun cycle theory and the dynamic theory, Lin Callis and Mark Schoeberl, reacted fiercely (Shell 1987). They claimed that the results were preliminary only and not a confirmation of the chemical explanation. 'Despite the number of public announcements, no clear link between manmade pollutants and ozone depletion over Antarctica has been established', Schoeberl said. As other observers have commented,

> The growing controversy about the cause of the ozone hole represented more than just differing scientific interpretations of existing data. It reflected the diverse instinctual responses among scientists and policy makers to the threat of large-scale ecological change. . . . A faith in nature's benevolence or, conversely, the conviction that the environment was highly vulnerable to manmade changes, could not help but influence the debate and directly contribute to the formulation of scientific theory
>
> (Cagin and Dray 1993: 291)

In November 1986, merely one month after their return from the NOZE expedition, a special issue of *Geophysical Research Letters* appeared, carrying over forty articles on the Antarctic *problématique*. Among them were numerous papers by authors who favoured a dynamic explanation. Since the authors appearing in this special issue wrote their papers independently of the NOZE researchers – both paradigmatically and geographically – on both sides the impression emerged that the other side would ignore one's own position. The dynamicists attacked the NOZE participants for having started their expedition with a prejudiced opinion. They wanted to prove the chemical theory, but did not succeed. However, neither did they succeed in disproving the dynamic theory (cf. Kerr 1987; Solomon 1987).

Indeed, there were many problems with the data. The best indicator in support of the chemical theory would have been the presence of chlorine monoxide in the stratosphere. The NOZE researchers tried to verify this through ground-based measurements but struggled with the task. The trustworthiness of the data was undermined when it turned out that their measurements of NO_2 were flawed (Roan 1989: 177; USAS 17). Another experiment tested NO_2 and $OClO$ concentrations. High NO_2 values would boost the sun cycle theory; high $OClO$ values would give credence to the chemical theory. The problem, however, was that $OClO$ is not an ozone depleting substance. It is created as a by-product in heterogeneous reactions that occur during the break-up of the two reservoirs hydrochloric acid and chlorine nitrate (see p. 37). High values of chlorine dioxide indicated only in a very indirect way that the chemical theory was right since they made the heterogeneous reactions plausible.

[Regarding Susan's ClO_2 measurements], we don't know what ClO_2 is anyway, it's a dead-end by-product, so we don't really know. There is a lot of confusion around that. The NOZE expedition was not very convincing in my opinion – Susan will tell you different, I am sure.

(USAS 17)

The Stony Brook team measured the very important radical, ClO, and found that there were two peaks in its vertical distribution, one up at the altitudes expected from the initial Rowland–Molina homogeneous reactions, and another much larger peak in the lower stratosphere. The difficulty with these data was partly that this group did not have very powerful computational facilities in Antarctica, and were unable to do the peak shape analysis effectively until they got home. While their publication came out before the 1987 expeditions, it was rather subsumed by the excitement of what looked to be a more convincing technique (see below) (personal communication with Rowland).

If one looks back at the 'hermeneutic triangle' presented in Chapter 1, one has to conclude that NOZE did not succeed in proving that the ozone hole was caused by CFCs (see Figure 1.4, p. 26).

The second Antarctic expedition, 1987

In this situation the proponents of the chemical theory felt the pressure rising. If it turned out that the Antarctic phenomenon was a natural phenomenon, the political process aiming at international regulations of CFCs would falter. Interestingly, both the decisive stage of the political process and the Antarctic experiments took place in the same time period (1986–87).

Large-scale experiments in general, including expeditions into the Antarctic and elsewhere, are called 'campaigns'. According to *Webster's Dictionary*, a campaign is an 'organized course of action for a particular purpose, esp. to arouse public interest; series of military operations in a definite area or for a particular objective'. In fact, the campaigns aroused public interest and the similarities with military operations are striking. The second expedition, particularly, was planned like a military action; military gear and personnel were used and the world public looked at the events. And, as in real war, the results of this campaign led to the victory of one side and the defeat of the other (Roan 1989). The men in charge at NASA saw the project as a miniature version of the Manhattan Project. As with the Manhattan Project, the challenge was to gather the world's best scientists in a unique effort to fight a life-threatening problem. The second expedition was to cost 10 million dollars and involved 150 scientists. Apart from ground and balloon measurements, *in situ* measurements in the stratosphere were scheduled. For this purpose, a modified version of the famous spy plane U-2 (the ER-2) was waiting to take off. Never before had a plane flown at this height under such adverse conditions (vortex, temperatures 85 degrees Celsius below zero and colder). The pilots who were ex-Vietnam combat pilots had to be persuaded to do the dangerous job. It was pointed out to them that the results of this mission were not only to be used in scientific research but would also help to solve an international political problem.

The programme manager of NASA had to find an airfield that was close to Antarctica. It was found in the South of Chile (Punta Arenas); the permission of the Chilean government had to be obtained and a runway had to be built in no time.

However, the scientific infrastructure also had to be in place. Jim Anderson, especially, was under great stress. Watson relied upon him and the highly accurate instrument he had built to measure chlorine monoxide. After the ambiguous results of the first expedition he feared another ambiguous result more than no result at all: 'There was no such thing as going in and coming back with some half-baked solution. . . . [That] would have been worse than no result at all' (cit. in Roan 1989: 187).

Anderson's team worked from March to May 1987 to adapt the ClO instrument to airborne conditions. The measurements, which were eventually taken in September, were successful; they can be described as *experimenta crucis*. However, the results would have not been so conclusive if Anderson's group had not established its status as the ultimate and undisputed specialists of *in situ* measurements in previous years.[23] Likewise, if the military-like organisation would have failed, the results would have been different (cf. American Geophysical Union 1989; NASA 1987).

> We had, of course, worked for nearly a decade to develop instruments to do *in situ* measurements in the stratosphere in parts per trillion. We've never flown one on an aeroplane before; there was a large learning process which we had to do very quickly. But spectroscopy kinetics, the heart of the method, we had forged under very controversial conditions for high-altitude measurements on balloons. So we've gone through a very significant period of technical growth on our own part within this research group. We could never have responded as quickly as we did to this question, and we still had a lot of homework to do after these flights. We had to refine the calibration, but they were [just] refinements. [They were] important refinements, but they did not strike at the heart of the fundamentals.
>
> (Personal communication with Anderson)

As a result of these flights a negative correlation between ozone and chlorine monoxide was established. The findings were plotted in a very convincing graph, showing both values as a function of latitude, i.e. as soon as the aircraft crossed 68 degrees South, ozone dropped dramatically and ClO rose exponentially (see Figure 3.5, p. 100).

Scepticism and trust: inclusion and exclusion

In this section I shall deal with two basic mechanisms of social action: scepticism and trust and their relevance in science. Following a widespread view (Merton, Popper, Luhmann), one could define science as that part of society in which scepticism reigns. The counter thesis holds that in science we can identify the pattern of group cohesion that is universal in social life. Elias and Scotson put it this way:

> The internal opinion of any group with a high degree of cohesion has profound influence upon its members as a regulating force of their sentiments and their conduct. If it is an established group, monopolistically reserving for its members the rewarding access to power resources and group charisma, this effect is particularly pronounced. . . . As competitive in-fighting of some kind . . . is a standing feature of cohesive groups, the lowering of a group member's ranking within the internal status order of the group weakens the member's ability to hold his or her own in the group's internal competition for power and status.
>
> (Elias and Scotson 1965: xxxix)

Several authors have emphasised the role of core groups in science. Such groups are small and tightly coupled and work on problems that are considered central, using innovative methods. Given this structure, it is mainly they themselves who define what the central problems are and which methods are to be considered innovative. They determine the further direction that a research field takes.

> Core sets funnel all of their competing scientists' ambitions and favoured alliances and produce scientifically certified knowledge at the end. These competing ambitions and alliances represent the influence or 'feedback' from the rest of the web of concepts and therefore the rest of our social institutions. . . . The core set 'launders' all these 'non-scientific' influences and 'non-scientific' debating tactics. It renders them invisible because, when the debate is over, all that is left is the conclusion that one result was replicable and one was not.
>
> (Collins 1985: 143f.)

In these groups there are many face-to-face interactions, personal contacts and exchange of unpublished research results. Recruitment to these insider groups often takes place via teacher–pupil relations. Trust and reputation circulate in core sets of scientific activity as they do in policy networks. This insight is important. While such a sociological analysis points out that social mechanisms like trust and credibility matter, scientists themselves unavoidably perceive consensus in abstract cognitive terms. Drawing on the earlier distinction between data and interpretation (see Chapter 1), scientists can only see data as influencing their scientific judgement. The scientific mindset does not allow that things other than facts and data have an influence on their verdict. This view is echoed by many social scientists. When analysing risk controversies, they often use terms like 'knowledge' and 'information' to describe the role of scientific experts on the policy process (see Breitmeier 1997; Litfin 1994).

Core sets of research are to be found at a few prestigious institutions and laboratories and are linked to other networks at prestigious institutions. Leadership personalities in these groups control the access to key resources of research such as labs, publication opportunities and finance (cf. Hagstrom 1965; Traweek 1988). They also decide in which direction the field will move, what the next 'hot' topics will be and where the boundary between science and non-science has to be drawn (Gieryn 1995; Jasanoff 1990).[24] The acknowledgement as an expert in the field of ozone research is done by the established core set of scientists. The following interview excerpts illustrate this:

> T. is a former vulcanologist who knows a little bit about the atmosphere, not very much. He was also the minister for industrial hazards in the Mitterand government in the beginning. He has some credibility, but he is not really doing research at all. He writes books on it.
>
> (USAS 41)[25]

> I think M. has no [reputation]. . . . These people do not show up in refereed journals. I made that statement about S. at one point. And his response was that he had written in *Science* and I knew about it because my rebuttal was in the same journal. That's sort of disingenuous on his part . . . because he knows that most people don't know that *Science* has two parts to it and that the letters aren't refereed. That is established by the fact that they have responses to them. . . . The real key is not whether the manuscript in question is called a Letter, but whether it is printed in a section of the journal which contains responses solicited from authors commented upon in the manuscript.
>
> (USAS 16)

> S. is now a lobbyist, no longer a scientist. His basic role is to ask questions. He uses the attitude of a sceptic. I never thought that he really understands this whole issue, but [he does] raise good questions.
>
> (USAS 41)

> There are also scientists, people who have done good work in the past, such as S. and E. But they have the wrong arguments. They do not publish in our journals. They do not get through with their arguments. Now one might say that this is the scientific Mafia which stops them but it really is too ridiculous.
>
> (GEAS 25)

As mentioned, WMO and UNEP started publishing a series of international reports from 1985 onwards. They are institutions, even obligatory passage points (Latour 1987) for policy makers and scientists alike. No scientist in the field of ozone research can afford to ignore this institution. Those who do not participate do not have a voice within the community (as one scientist told me, 'the sceptics don't go to these meetings'). It appears that one interesting aspect of boundary work is that the mixing of scientific and political statements is not prohibited in any case; it all depends if one is recognised as a scientist. If one succeeds in distinguishing one's own position as truly scientific, those of competing claims makers as politically biased and non-scientific, then one has made a double achievement with a single step, i.e. separated scientists from non-scientists and legitimate from illegitimate partisanship.

Core sets and boundary work have another implication as well. Scientists on the research front never have absolute certainty about the results of their research. They have to trust the data that they mobilise to foster their own position. If a scientific speciality agrees on specific methods, explanations and techniques, those who pose sceptical questions at the 'wrong time' remain outside the core set (cf. Shapin 1994). Within the community, there seems to be a window of opportunity for making sceptical statements that touch upon the fundamentals. One scientist stated quite categorically that it is their task to be sceptical: 'All of us reacted sceptically when they brought this up there. That's our job to be sceptics. You have to prove it' (USAS 45). After the community has embarked upon a certain path, it does not like to hear sceptical arguments any more. A closure process takes place. This process unfolds its dynamics only where the scientists within a research community cooperate. In cases where such a community does not exist, scepticism has more leeway. If scientists deem some unconvincing result as worth mentioning, the open, sceptical reaction is likely. A somewhat embittered participant of the ozone controversy of the 1970s, who has withdrawn from the dispute, confirms this when he says:

> Unconsciously, they formed a tribe that has a mutual self-interest in sustaining the ozone crusade. They do not like to have critical comments made from outside, it's understandable, it's human. The scientists formed a community, they were ably led and that was it. They all agreed. Nobody would want to disagree. If you had been at some of those meetings you would know it. I will never forget the hostility if you got up and suggested anything that was contrary to the message.
>
> (UKAS 42)

From inside, this is expressed as follows:

> The people who I knew who I would have thrown out of the review process were the perennial troublemakers. In other words, they were given four chances to shoot it down and they each try, and after four times you say: Do a better job! People should not bring you disproofs that have obvious flaws in them all the time.
>
> (USAS 15)

As is the case with any closure process, the insiders get closer together, pushing marginal people outside. The insider who was just quoted as saying that it was the job of scientists to be sceptical, explains:

> It is certainly true that there is a paradigm in our field as to how things work,

and things which walk outside of that paradigm get a rougher road in reviewing. I can imagine kinds of papers that I could write that would have a hard
time.

(USAS 45)

How important it was to reach agreement can be gathered from the problems that
the field faced:

You will always have this problem when you make scientific measurements. It
is rare that any two different methods will come to the same number. When
you are within 10 per cent, you are pretty good. Now we try to get at 1 per
cent changes in ozone, so there is a lot of fights about what's more accurate.

(USAS 17)

As a director of a lab which has one of the world's great responsibilities to get
good numbers on global warming, we are viciously critical of things like that
in our own models, because everybody is viciously critical of them for the
simple reason that they don't want to change policy on the basis of dubious
models and the draconian implications of greenhouse warming. So if we
subjected the ozone models to the same rigor as in the greenhouse debate,
there is a certain fuzziness remaining, that is very understandable given the
degree of difficulty of the science.

(USAS 11)

Various sociologists of science have remarked that in cases where 'objective' or
'scientific' tests are not available to check the quality of experimental data, scientists are making free use of non-scientific criteria, such as trust in the honesty of
an experimentor, the size or prestige of a lab, even personal characteristics like
nationality or professional specialisation (Collins 1985; Holton 1994; Shapin
1994).

Trust in the work of others

The discovery of the ozone hole provides rich illustration for the thesis that trust is
a central notion of scientific work. After scientists of the British Antarctic Survey
(BAS) had published their article in *Nature*, the international research community
was puzzled. What is the BAS? Who is Farman? Are these measurements reliable?
Ralph Cicerone was quoted as saying, 'The BAS is not a household word. At the
time, most of us had never heard of it, had no idea whether these people did good
work. You couldn't automatically give credence to the work' (cit. in Roan 1989:
129). Another researcher told me that he initially believed that Farman's data
were faulty measurements. This was confirmed by other interviewees: 'When I
heard about the discovery of the Antarctic ozone hole, I thought it must have
been very bad measurements' (USAS 11).

The measurements taken by BAS were one-point measurements where ozone concentrations were measured above Halley Bay station.[26] Many questioned the legitimacy of deriving a dramatic statement from these results. At any rate, NASA's satellite did not confirm the BAS measurements at the time. At a scientific meeting in July 1985 at Les Diablerets, Switzerland, Rowland asked about the British Antarctic Survey. There was a near total absence of information – including Rowland's own – except the content of the paper in *Nature*. The Les Diablerets meeting was the final preparatory meeting for the NASA/WMO report published in 1986, and there were some 100 to 150 atmospheric scientists there. However, the queries and concerns about Farman and the BAS were private, and not part of the general discussions. Two or three weeks after Les Diablerets, Rowland then travelled to a meeting in Honolulu where he met Brian Gardiner.

> With a Xerox copy of the British paper in his briefcase, Rowland left Switzerland for Hawaii where he met Brian Gardiner, one of the co-authors of the British paper. Rowland was impressed with Gardiner and became completely convinced of the validity of the British research.
>
> (Roan 1989: 131)

> I had thought it highly probable going to Les Diablerets that the British Antarctic Survey report was correct, but from Hawaii on, I was quite convinced that the high Antarctic losses were real.
>
> (Personal communication with Rowland)

Thus Rowland was one of the few scientists of the established community of atmospheric scientists who was convinced in the summer of 1985 of the trustworthiness of the British data. At the time, three important scientific conferences took place. Les Diablerets (July 1985) was followed by the one in Hawaii (late July) and one in Salzburg (August). Rowland returned to Europe for the Salzburg meeting in which Don Heath of NASA showed 'ozone hole' slides (they had also been shown in a meeting in Prague a week or two earlier). This meeting was much smaller, and Heath's work was obviously seen by all. The end result of that summer was that Rowland was quite convinced of the complementary significance of the Farman work and the NASA satellite work, and proceeded to talk that autumn about the 'Antarctic ozone hole' to Walter Sullivan of the *New York Times*, and in meetings that winter.

Decisive for the majority of scientists was the re-analysis of NASA's satellite data, which confirmed the BAS findings. The results of this re-analysis were published on 28 August 1986 in *Nature* (Stolarski *et al.* 1986; the research community knew one year earlier about the low Antarctic ozone values). Without any doubt, this publication was an embarrassment for NASA which lasted for quite some time. A responsible NASA scientist wrote on the matter:

> Unfortunately, everyone 'knows' that NASA did not discover the ozone hole because the low values were 'thrown out by the computer code'. This myth

was the result of a statement made by one of my colleagues in reply to a question during an interview. . . . He was not directly involved in ozone processing at the time and his answer was *not* correct.

(McPeters, cit. in Pukelsheim 1990: 541; original emphasis)

When I conducted my interviews at NASA, the scientist who did the re-analysis of the NIMBUS database (the same person who is referred to in the above quote?) confirmed that a computer algorithm had flagged the low values. Be that as it may, one can only agree with NASA's concern that 'the myth that our computer code "threw out the data" is unfortunately very hard to correct without appearing defensive' (McPeters cit. in Pukelsheim 1990: 541).

Trust in one's own work

Merton reminded the sociologists that self-confidence is an important requirement for successful scientific work. He says about the successful scientists: 'They exhibit a great capacity to tolerate frustration in their work, absorbing repeated failures without manifest psychological damage' (Merton 1973c: 453).

Many of us will have experienced the uncertainty that surrounds a talk or a paper that strays too far from what the mainstream is ready to accept. The same uncertainty emerges when making statements that cannot be proven readily – but of which one is convinced. 'Belief is not knowledge'; we all seem to remember this admonition from schooldays. In order to foster innovation, one needs a counterweight against the pressures of conservative and established bodies of knowledge. Trust in one's own work is a precondition. Without self-confidence it is unlikely to resist the pressure of the environment, above all the pressure that comes from sceptical colleagues. The more central the old body of knowledge has been in a particular field, the greater the pressure. The pressure also increases if the context of application brings unpleasant results for powerful actors, or if the 'deviant' researcher leaves the confines of academic research and enters the public fray (or is pushed into it). The Hollywood movie *The Insider* nicely captures this problem as a lawyer points out to those who want to convey a critical message about dubious practices of the tobacco industry that 'the greater the truth, the greater the damage'. As we shall see in Chapters 4 and 5, in many cases the chemical industry reacted fiercely to scientific results which posited CFCs as a cause for great concern. In such situations, an innovative researcher needs the backing of a community in which she can discuss the results of her work and in which she finds allies for the battle in public. Those researchers who do not belong to such a community find themselves in a precarious situation. The person who discovered the ozone hole seems to have been in such a situation. The support in the academic scene was not enthusiastic when the referees wrote the reports on the 1985 paper. One referee's report stated that the result in the paper was 'quite impossible, but if it is true it is actually quite important, better publish it'. The other cast doubt on the link created between the Antarctic phenomenon and CFCs. Yet what is more, ICI, one of the big CFC producers, had obtained a copy of the article before it was published and tried to pressurise Farman:

> ICI rang me up, they had a copy of the paper long before it was published. And ICI said: You must not publish that, it is not science! And I said: No, it is not science. But I am going to publish it. It does not prove anything, of course it doesn't. How can you prove such things at this stage? I feel that this is the first real effect of CFCs. You can see vaguely how it can happen.[27]

The article was published in the section 'Letters to *Nature*'.

Implicit knowledge

Polanyi noted that the consensus about what counts as science and who is a scientist is established by repetition and confirmation of experiments and results. By so doing, however, one would only confirm the belief in the consensus without having an independent source of judgement:

> But the affirmation of this supposed fact is actually but another manner of expressing our adherence to the consensus in question. For we never do repeat any appreciable part of the observations of science. And besides, we know perfectly well that if we tried to do this and failed (as we mostly would), we would quite rightly ascribe our failure to our lack of skill.
> (Polanyi 1958: 217)

To be sure, during the CFC controversy several measurements have been repeated in labs and in the atmosphere. In some instances this was not possible immediately, for example, the discovery of the ozone hole and the airborne measurements in September 1987 over Antarctica. In the first instance the repetition followed some months later by means of another measurement method, in the second instance they were repeated later in time. The discovery of the ozone hole by Dobson instruments underlines the central importance of implicit knowledge, even in seemingly trivial activities such as the reading of values from an instrument.[28] The problem of reliability of time series data has been illustrated by the work of the Ozone Trends Panel. I have already addressed the work of the OTP and the problems of standardisation of Dobson stations. Here I focus upon the aspect of implicit knowledge in handling the Dobson instruments.

> If you are running a Dobson properly, you are making 100 measurements in order to get five ozone measurements. All the other ones you are taking is to convince yourself that there is consistency between all the different methods. The absolute method is to look at the sun. But then you also do it on the cloudy sky and various other things as well. And then you've got to convince yourself that all these other observations are all related back solidly to one thing you are sure of which is the sun observation. You just have to work very

hard to get your sun observations right. You have to work even harder to go through the whole chain of weather conditions. Then you have to compare them and find the error bars. In the end you can even go out on a cloudy sky and still take a reading and convince yourself that you know how to relate that to the direct sun reading which you would have got.

(UKAS 44)

So, taking measurements with a Dobson is no trivial activity, as this expert tells us.[29] Originally, only qualified persons had been given this task. However, this kind of activity gradually lost more and more of its appeal. As a direct result, the data lost quality. Bad measurements were the consequence, and they were not isolated cases.

[The Dobson] got pushed down and down the chain of expertise until it has been done by young people who did not know what the hell they were doing. In L., ten years ago, the man would go out of the hut, switch it on, turn the dial, read the number of the dial, pick up the telephone [and] give that number to B. He would have no idea what it meant in terms of ozone himself. Someone in B. would turn it into an ozone value, and sooner or later it would appear in the red books in the world series. That's no way to run a system.

(UKAS 44)

The BAS had trained young people to sit in the ice desert, thousands of miles from home, only to operate an instrument. Every year two people took turns in the service at the station. In the late 1970s, when measurements started to arrive that were outside the expected range, the BAS tried to find out the reasons. The scientists of the BAS thus asked themselves: Can we trust the data that we get from our station in Antarctica? If the measurements were bad, there could only be two reasons: the technicians or the instruments.

You make sure your young man does that properly, I mean you are not looking over his shoulders. Taking the observations yourself would be a different thing but you're training young men, you're sending them 10,000 miles away, you know they can do silly things from time to time. And eventually you start worrying about the instrument itself, they are both fairly old instruments.

(UKAS 44)

In view of these imponderables, the BAS did not publish their results but waited until they were certain that they had not analysed faulty measurements.

Rhetorics and inscriptions

Symbolic representation

Even before the British team published their results, a Japanese group of researchers had found abnormal ozone values in the Antarctic. However, because they were isolated from the rest of the international community of atmospheric researchers, they did not and could not present their findings in a way that would have alarmed the world public. Their data contained only a time series of 11 months. They presented them in 1984 during a poster session at an international scientific ozone conference in Greece, one year before the publication of Farman's results.

The results were not published in a major journal but in a rather obscure outlet (Chubachi 1984). It seems no exaggeration to say that they did not realise what they were measuring. They stressed the anomaly of an exceptional high value in October (which occurred after the concentration had dropped to 240 DU). In other words: their framing did not catch the attention of their colleagues nor of the world public.

> The Japanese were measuring ozone in their station in Antarctica. And they found abnormal ozone levels. They reported that in a meeting in Thessaloniki. They had a poster, and you know how people look at posters. Nobody really paid attention. They had abnormal values, so what?
>
> (USAS 41)

Since the Japanese researchers were not familiar with the discussion of the core set of atmospheric scientists, they were not able to frame their results in a more appropriate way. The BAS was different. The team knew about the key issues in ozone research. They were aware that they had found something explosive. Thus, they kept their data secret for a long time in order to avoid sounding a false alarm (Roan 1989: 125). Farman's article in *Nature* contained a very suggestive figure which displays the decreasing ozone values over Antarctica together with an increase in CFC concentrations in the Southern hemisphere (see Figure 2.2, p. 43).

As we have seen, this put some referees off. They thought this was mere speculation; however, the article was eventually accepted for publication.

> Farman made statements, also in the press, that it must be CFCs. These were rather convincing to the public because Farman had this appealing plot which shows ozone decline and CFC increase in the same graph with an appropriate scaling so that the two match. To the scientists this indicated a possibility that the two could be related but the evidence was quite weak at that time.
>
> (USAS 38)

Another atmospheric scientist also bemoans the speculative link of ozone and CFC concentrations:

> [T]hat figure where he suggests a correlation between growth in CFCs and drop in ozone . . . was scientifically not justified. You may make an equally justified plot between the Dow Jones industrial index and the ozone hole. If you have something going up and something going down then you can always slide the scales and it will look like a correlation, but there is nothing scientific about it.
>
> (USAS 30)

Despite all of this, the BAS succeeded in getting the attention of the core group of ozone researchers, and also of the world public.

In summary, one can state that trust in the quality of researchers and therefore their scientific data may accelerate the discussion process. This happened quite frequently during the ozone controversy since research results were made available before they entered the pre-print stage, in other words, before they had gone through the review process.

> The scientists were much less reticent to discuss their results prior to publication. The publications themselves still went through peer review in the normal way, but most of the active scientists knew what the new results were long before they appeared in the literature. This . . . global basis of this interchange was relatively unusual in the mid-1970s.
>
> (USAS 16)

These examples highlight the role of core sets in which one profits from the trust of others. They also illustrate what it means to be an outsider (or a *nobody*), whose results are not taken into account. In the case of the ozone controversy, the discussion could be accelerated by circumventing formal rules – something which happens in core sets all the time.

The following three sections focus upon the production of symbolic resources that were decisive for the result of the controversy. I draw the attention to visual and rhetorical means of representing scientific claims.

Images and metaphors

As Bruno Latour has pointed out, visualisation is essential to establish scientific knowledge claims. But he is also aware of the danger that we might invest symbols and signs with a power reminiscent of mysticism:

> We must admit that when talking of images and print it is easy to shift from the most powerful explanation to one that is trivial and reveals only marginal aspects of the phenomena for which we want to account. [They] . . . may explain almost everything or almost nothing. . . . My contention is that writing and imaging cannot by themselves explain the changes in our scientific societies. . . . We need to look at the way in which someone

convinces someone else to take up a statement, to pass it along, to make it more of a fact.

(Latour 1990: 23–4)

Latour does not 'find all explanations in terms of inscription convincing but only those that help us understand how the mobilisation and mustering of new resources is achieved'. Therefore, one has to invent objects that are mobile but also immutable, presentable, readable and combinable with one another. He coins the term 'immobile mobiles' for this achievement. Inscriptions are mobile but immutable when they move; they are two-dimensional and their scale can be modified and recombined at will. 'The phenomenon we are dealing with is *not* inscription *per se*, but the *cascade* of ever simplified inscriptions that allow harder facts to be produced at greater cost' (Latour 1990: 40–6).

Simon Schaffer questioned Latour's concept of 'immutable mobiles': 'Despite recent insistence on the immutability of mobile inscriptions, pictures were always embedded in rather complex technologies that were not easy to translate, and their evident meaning relied on interpretative conventions that were by no means robust.' He thus insists that every representation depends on craft and local contexts (Schaffer 1998). While this seems to be an issue that can only be decided on empirical grounds and not in any a priori way (there may be cases where immutable mobiles in Latour's sense emerge and thereby close off a debate), Schaffer also makes a more theoretical point which deserves our attention. Drawing on the work of Ludwik Fleck, he states that:

> the picture of science as an esoteric zone, where facts are made, and an exoteric zone, where they are consumed, is wrong. . . . Facts and representations are formed in the spaces between these zones. Facts become robust by drawing both on esoteric private work, produced in labs, observatories, and technical institutions, and on the general strictures of popular culture.

(Schaffer 1998: 221)

Hilgartner (1990) captures this thought in a model of an 'upstream'–'downstream' continuum and claims that 'as scientific knowledge spreads, there is a strong bias toward simplification (that is, shorter, less technical, less detailed representations)'. The question therefore is: 'is the particular transformation "misleading" (and therefore blameworthy)?' (Hilgartner 1990: 529).

The pre-language element of pictures can persuade the beholder instantly, as if he was struck by lightning, something that would not be possible through argument. The beholder may convince himself at one glance that a statement is valid. In fact, particular diagrams and visualisations were enormously powerful in the CFC–ozone controversy. Investments in instruments and scientific infrastructure seem to affect the status of the results thus obtained. Scientists have more trust in results produced by new, expensive instruments than by old and cheap instruments.

'Smoking gun'

The anti-correlation between ozone and chlorine monoxide that had been measured by Anderson's team in August and September 1987 during flights into the
Antarctic vortex has been coined as 'smoking gun'. This notion, which seems to
stem from the Far West and which was often used during the Watergate scandal,
denotes a strong piece of evidence in a trial, without being the direct proof of
guilt. It is something that will probably be considered as a cause on closer examination. During the Watergate affair the secret tape recordings were a strong piece
of evidence that President Nixon did not tell the truth (*Webster's* dates the notion to
1974). My American interviewees used the 'smoking gun' metaphor for a particular scientific result, namely the measurements taken by Jim Anderson and his
team aboard the ER-2 (see Figure 3.5). The publication of these data (Anderson
et al. 1989: 11479) contains the following summary:

> While the existence of anti-correlation between two variables does not itself
> prove a causal relation, we present a case here for the observed evolution of a
> system from (1) an initial condition exhibiting dramatically enhanced (x500)
> ClO mixing ratios . . . to (2) a situation in which the dramatic increase in ClO
> recorded on surfaces of fixed potential temperatures was spatially coincident
> with a precipitous decrease in ozone mixing ratio.

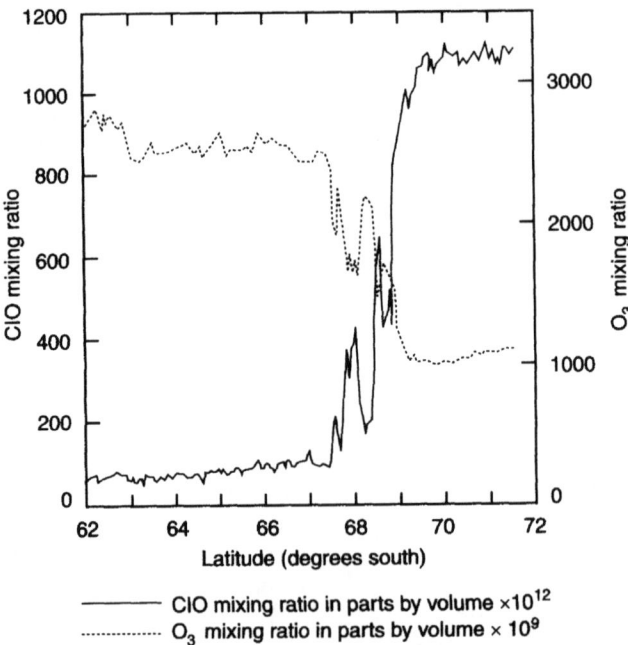

Figure 3.5 Airborne measurement of ozone and Chlorine monoxide, 62°S to 72°S
Source: Anderson *et al.* 1989: 11475.

Anderson *et al.* had two pieces of evidence: a negative correlation between ClO and ozone and the correspondence of observed kinetic reactions with theoretical predictions. Taking both together, they conclude that CFCs have caused ozone destruction. It is remarkable that they do not employ the notion of proof; instead they use the notion of making a case: 'When taken with an analysis of the kinetics of ozone destruction by halogen radicals, this constitutes the case linking ... chlorofluorcarbons at the surface[,] to the destruction of ozone within the Antarctic polar vortex' (Anderson *et al.* 1989: 11479).

The scientists were aware of the fact that they could not provide a final proof. In the meanwhile, the 'smoking gun' in this case is seen as much as proof as it was in the case of Nixon's political fate. Scientific results seem to become more robust the older they get (Collins 1985: 145, fn. 15). When asking atmospheric scientists, I detected an interesting ambivalence on the issue of proof. Asked about the status of their results, they were in general very cautious about using the term proof:

> Getting absolute proof is nearly impossible. But we do have a ... list of levels of what's known ... what we know less, what's plausible, and what isn't and what we eliminate. That's really where the state of the science is. That is probably the best we can ever do, we can just change what's in what portion of that list. But we can never get an absolute proof. The atmosphere is too large and too complex.
>
> (USAS 47)

> In the environmental sciences there is no proof in a mathematician's sense. The word proof to me does largely reside in mathematics or logic where it either is or it isn't. In the experimental sciences and certainly the environmental sciences, you build only a stronger and stronger circumstantial case. The reason is you cannot measure everything everywhere in an environmental issue. But you try to be clever enough to build, to measure, to theorise and test in what you believe to be the crucial places.
>
> (USAS 46)

> Is it a proof or an established fact? I would say that most of the stuff we deal with it is hard to think of as a proof because how can you rule out all of the causes?
>
> (USAS 15)

Anderson and his colleagues avoided the term proof for the airborne results as well. Over the years, however, the results gained the status of proofs, as the following statement makes clear: 'Once we had observed through Jim Anderson that ClO was in the stratosphere, that again said, boy, now we've actually got the proof that there is the radical [ClO] that the model predicts destroys ozone' (USAS 39).

The invention of the ozone hole

Joe Farman and his collaborators discovered the ozone hole, Sherry Rowland invented it. Farman reported about abnormal seasonal ozone values over one ground-based station. He did not use a metaphor to describe the results and did not use the term ozone hole. And he could not have used it since his data contained a time series at one geographical point, illustrated in a line diagram (Figure 3.5). The title of the article was: 'Large losses of total ozone in Antarctica reveal seasonal ClO_x/NO_x Interaction.' The metaphor did present itself only when the low Antarctic values were displayed in a different graphical manner. During scientific meetings in 1985 in Prague and Salzburg, Don Heath who had built the TOMS-instrument for the NIMBUS-satellite and had been involved in data retrieval (see above), showed some colour slides with satellite data reaching from 1979 to 1983. This bird's eye view showed global ozone values over the Southern hemisphere, i.e. the low Antarctic values in context. Only in such a form of representation one was able to see a 'hole' (Figure 3.6).

Having seen the slides in Salzburg, Rowland obtained them from Heath and showed them at a talk in November 1985 at the University of Maryland. During this occasion, he used the term 'ozone hole' – this was probably the first time the

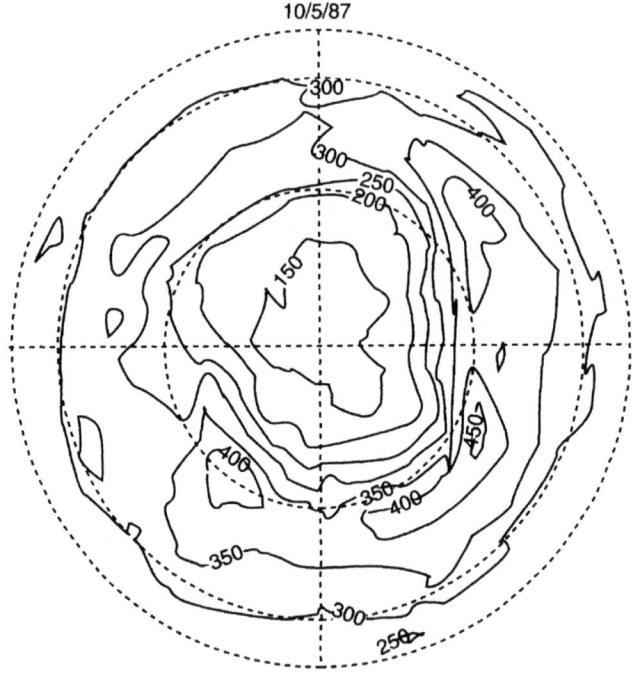

Figure 3.6 Dramatically low ozone over Antarctica (50% below normal values, in Dobson Units). Data from NASA satellite Nimbus-7, taken on 5 October 1987

Source: WMO 1988: 684.

term was ever used. Before the seminar started, he sent out a press release and phoned Walter Sullivan of the *New York Times* who ran an article the next day, using the term 'hole'. It also had an illustration similar to the one in Figure 3.6. The student paper at the University of Maryland reported on 8 November 1985 under the headline 'Scientist warns about "hole" in ozone layer'. Rowland is quoted as saying 'We've used up our margin of safety and we've used it up frivolously'. He demanded an immediate ban of all CFCs in the USA and worldwide with the exception of medical applications.

Rowland transported this powerful metaphor directly to the world public. In scientific publications it took longer before the term was accepted. The paper by Stolarski *et al.* (1986) which contained the first 'official' confirmation of Farman's findings was not allowed to use the term in its title, as the authors had intended. Its final title was 'Nimbus 7 satellite measurements of the springtime Antarctic ozone decrease'.

> When we submitted the first data on that 1986 *Nature* paper, we used the term 'ozone hole' in the title and one of the referees objected to it. So we changed it. It is one of those terms where all the scientists said: Gee, this is not a very good term, but once it had been said, it was inevitable. It was such a simple description, it's a code word that means 'that phenomenon down there'.
>
> (Personal communication with Stolarski)

Shortly after Stolarski's publication, Crutzen and Arnold published their explanation of the ozone hole in *Nature*, using the term 'ozone hole' in the title (Crutzen and Arnold 1986). However, it took some time before the term became accepted by the community of atmospheric scientists. This can be seen by analys-ing the frequency of the term in editorials, letters and notes versus the frequency of the term in published research results. During 1986 and 1987 there were very few articles (research results), but many editorials and letters (Table 3.1) using the term. After 1989, this difference was reversed: now there are more articles con-taining the term. This seems to suggest that referees only reluctantly accepted the term when the journalists and editors did not see a big problem in using it. After it was accepted in the scientific community, it was used as a common term. Thus, the metaphor of the ozone hole has not been designed for special public use. This contradicts dominant views about popularisation in which scientists are assumed to possess 'pure knowledge' which is then polluted when it enters the public sphere (see also Bucchi 1998; Hilgartner 1990; Irwin and Wynne 1996).

Table 3.1 shows the difference between scientific articles on the one hand and research notes, editorials and letters on the other. One can see that there were only two articles in 1986, compared to nine editorials, letters and notes. Only after 1987 did the term ozone hole acquire wider currency. This indicates that there was an uneven development within the scientific community. In fact, some early advocates were convinced throughout the controversy that the problem would not go away and that in the face of doubt precautionary policies were in order. Other parts of the community preferred a more passive approach, waiting for more

Table 3.1 'Ozone hole' in scientific journals

Year	1985	1986	1987	1988	1989	1990	1991	1992	1993	1994	1995	1996	1997
Articles (in title only)	0	2	7	11	12	8	14	7	6	2	10	10	13
Editorials, letters, notes (in title only)	0	9	10	12	13	3	6	12	5	4	4	2	1

Source: SCI.

evidence and giving industry the benefit of the doubt. However, advocates succeeded in alarming the public even though parts of the scientific establishment (such as important journals) were opposed to their rhetoric.

The symbolic meaning of the term ozone hole can be grasped through an analysis of the semantic field in which it is located. From the beginning of the controversy in 1974, the metaphor of 'protecting the ozone layer' had been established in public discourse. Like every metaphor it combines a notion with a picture and an analogy (Black 1961; Lakoff and Johnson 1980). In this case the picture was one of the globe with an outer protective layer that might get hurt. It was a matter of debate how severe the wounds inflicted by industrial chemicals would be. However, both sides saw the problem in terms of wounding, not complete vanishing (dying). In another picture, taking the layer in terms of clothing, there was a tissue becoming threadbare. Both sides had different opinions about how far we might go. We could not buy a new coat, so much was clear. Rowland's conceptual innovation amounts to a change in the root metaphor which revolutionised the whole perception of the problem. The picture of a hole evokes a balloon that explodes or loses its air (the German news magazine *Der Spiegel* (49, 1987) created the metaphor of a 'Leak in spaceship earth'). The much shorter time dimensions provided a dramatic element that is well captured by this metaphor: it is no longer an ozone loss of 10 or 20 per cent in 100 years but of 50 per cent *now*. This sudden loss happens only in one geographical region; but it may well be that this is only the prelude to a global loss of the ozone layer. The metaphor raises the anxiety that in the coming years the hole will grow bigger until it eventually covers the whole planet – much like a wound eats away more and more of an infected organism.

4 Country comparison

The case of the United States

Political context

The CFC–ozone controversy took off in the United States in a specific historical and political context. This country has one of the longest histories of concern for the environment; several environmental issues had been raised and institutions had been created before similar developments took root elsewhere. The beginning of American environmental politics and policies are usually linked to two land-marks: the publication of Rachel Carson's book *Silent Spring* in 1962, which sounded the alarm on pesticides, and the Earth Day in 1970 (Hays 1987). In the mid-1960s, several pieces of legislation were passed: the Wilderness Act in 1964, the first Clean Water Act in 1965, the Clean Air Act in 1967, and the Wild and Scenic Rivers Act in 1968. Since then, major environmental groups have emerged, mainly based in Washington and relying on legislative and litigative initiatives focused on federal government (Dowie 1995). In 1969 the National Environmental Policy Act (NEPA) was passed, leading in turn to the founding of the Council on Environmental Quality (CEQ). The latter was intended to facilitate the president's drawing up of an annual environmental report. In 1970 President Nixon recommended the formation of an Environmental Protection Agency (EPA) to Congress. Its primary task would be pollution control. Nixon accepted that the environment is an interdependent system, so the EPA was not to be organised according to environmental media (air, water, soil), but, rather, functionally. This meant that it identified harmful chemicals, followed them through the entire ecological chain, and determined the chemicals' effects and the interactions between them, in order to determine ultimately where in the ecological chain it was most sensible to intervene (see Hays 1987; Marcus 1991).

The implementation of this ambitious goal was not successful, the main reason being that the EPA was under a great deal of pressure. The first head of this agency, William Ruckelshaus, feared that the difficult question of the agency's organisation would prevent it from being able to deal sufficiently with the urgent problems of day-to-day politics. The functional method of organisation could thus only be arrived at step by step. The EPA was also under pressure from the demands of the environmentalists, who expected that rapid measures would be implemented and polluters taken to court. The Clean Air Act, moreover, mandated the establishment of national environmental standards. Finally, the White House wanted to see progress in the form of cost–benefit analyses. Ruckelshaus compared his task with that of an athlete who runs the 100 metres while having his appendix removed (cit. in Marcus 1991: 22).

Long periods of the 1970s were characterised by the dominance of two topics: the oil crisis and toxic chemicals. The oil crisis of Winter 1973–74 brought the question of energy supply onto the agenda and gave more influence to the concerns of industry, which had previously been curbed by environmental programmes.

During 1976 and 1977 there was a brief flirtation between the newly elected President Carter and the environmental movement. Various legislative initiatives were supported by the spokespeople of environmental groups. In autumn 1977, however, the economic advisers in the White House gained the upper hand. Their agenda included the hobbling of further environmental laws, which in their eyes were slowing economic growth. This group's influence increased during the economic crisis of 1979. The electoral victory of Ronald Reagan in the following year finally brought a frontal attack onto all the environmental programmes of the previous 20 years.

Scientific context

In 1971 congressional hearings were held to discuss the atmospheric effects of Boeing's planned supersonic transport (SST). The main focus was the role of nitrogen oxides (NO_x) emitted in the operation of these aircraft. The effects of nitrogen oxides on the ozone layer were discovered by two researchers working independently of one another (Crutzen 1970; Johnston 1971). Without this debate about the assessment of technological consequences, the ozone layer would probably never have arrived on the agenda.

As I discussed in Chapter 3, two other researchers discovered the ozone destroy-ing effects of chlorine, an effect which hitherto had been completely unknown (Stolarski and Cicerone 1974). Volcanoes and emissions from the planned space shuttles were initially posited as sources of stratospheric chlorine, although at first one could only speculate about the quantities emitted. Molina and Rowland's innovation was the identification of a source (chlorofluorocarbons) that could bring considerable amounts of active chlorine into the stratosphere. They were made aware of this possibility by Lovelock's published measurements, which he had taken with the electron chromatograph he had constructed (Lovelock *et al.* 1973). With this instrument it had become feasible to measure gases in concentra-tions of billionth parts (10^{-12}) and beyond. Rowland knew enough about photo-chemistry to consider the possibility that CFCs remained stable in the troposphere (and were therefore used by meteorologists as 'tracers'), but could be destroyed in the stratosphere by UV light. In this process, he speculated that active chlorine would be released and the catalytic chain reaction discovered by Stolarksi and Cicerone could begin.

This course of historical events suggests that the attention of the scientists, the scientific community and the wider public was attracted by means of a chain of specific circumstances. First, it was important that Molina and Rowland and Cicerone and Stolarski established contact with each other. Interestingly, this contact was initiated by Hal Johnson, who had previous experience in dealing with the public and politicians from the SST debate. When Rowland and Molina first told him of their hypothesis, Johnson presciently asked: 'Are you ready for the heat?' (cit. in Roan 1989: 19).[1]

In fact, the contextual conditions were beneficial: there was public attention given to the ozone layer and a handful of scientists were ready to play a public role. As will be shown, these speakers form the core around which further controversies, actors and resources settle.

Policy networks in action

Actors

The personal nucleus of the advocacy coalition was formed by three of the four scientists who had produced the crucial scientific publications on ozone destruction by chlorine and/or CFCs. In the subsequent months they were joined by representatives of the Natural Resources Defense Council (NRDC), the Council on Environmental Quality (CEQ), the Ad hoc Federal Interagency Task Force on the Inadvertent Modification of the Stratosphere (IMOS), the National Academy of Sciences (NAS), the Environmental Protection Agency (EPA), and the Consumer Product and Safety Commission (CPSC), as well as science writers from influential newspapers.

The nucleus of the counter-alliance was the Manufacturing Chemists Association (MCA), particularly the Du Pont company. The speaker for the counter-alliance was the head of Du Pont's Freon section, Raymond McCarty (Freon was Du Pont's trade name for CFCs). The counter-alliance was joined by several scientists and publishing organs with ties to industry. The scientists from whom the counter-alliance received support did not form a homogeneous group. First of all, there was the British scientist Lovelock, who enjoyed a high standing due to his early concentration measurements. He was put off by the 'overselling/preaching' tactics of the American scientists:

> Some scientists are worried about certain gases used in aerosols. . . . This is one of the more plausible of the doomswatch theories but it needs to be proved. The Americans tend to get into a wonderful state of panic over things like this . . . I think we need a bit of British caution on this.
>
> (Lovelock, cit. in Bastian 1982: 171)

He stated that the 'American reaction to the problem means that scientific arguments no longer count. The impact of CFCs on the ozone layer has been grossly overestimated.' Moreover, he thought that there was 'inadequate medical evidence to support the belief that higher levels of ultraviolet radiation would necessarily lead to higher incidences of skin cancer' (*New Scientist*, 19 June 1975: 643).

Nevertheless, in a letter to *Nature*, Lovelock conceded that there could hardly be any doubt that unrestricted CFC emissions could have damaging and long-lasting repercussions. He had, however, a different view of what was to be considered a dangerous level and when it would be reached (*Nature*, vol. 258, 25 December 1975: 776).

Richard Scorer was engaged as a meteorologist at Imperial College (University of London) and was invited to make a propaganda tour of the USA by the CFC producers' interest group, MCA. His reputation was the lowest of the scientists introduced here. He had undertaken no research of his own in the field and was considered by many to be the CFC producers' hired gun.[2]

> The other scientists who were against regulations were clearly paid by various parts of industry. Like Richard Scorer from England, but he was not respected. McElroy was different, after all he was a Harvard professor, he was very famous, and later, after 1986, he became a supporter.
>
> (USAS 5)

At the very beginning, the well-known atmospheric scientist and Harvard Professor Michael McElroy had taken a similar position to Rowland's. In a newspaper interview he expressed the opinion that the burden of proof lay with the opponents of the CFC–ozone theory, 'because we can't wait to check out the full theory. It would take too long. The effects of Freon are too damaging' (*Detroit News*, Sunday Magazine, 19 January 1975). Shortly thereafter he spoke more cautiously, stressing that above all more time (one to three years) was needed for additional research. This, then, was also the lowest common denominator uniting industry and *these* scientists. The coupling between industry and the three scientists mentioned was relatively loose; the scientists wanted more time for research and downplayed the problem to various degrees.

Resources

As the controversy began, industry was unable to provide any scientific data to support its position. Although the Manufacturing Chemists Association had already established a research programme in 1972 to clarify what was happening with CFC emissions, they were surprised by the Molina–Rowland hypothesis. Their strategy thus consisted of emphasising the hypothetical character of Molina and Rowland's calculations, relativising the hypothesis by means of a counter-hypothesis ('no danger to the ozone layer') and suggesting that the CFC critics themselves also had no data. This argument was partially laid to rest by various laboratory and field experiments. Both substances, CFC 11 and 12, were measured in specific concentrations at different altitudes, corresponding to the theoretical expectations. It should be noted that Lovelock carried out the earliest measurements of CFCs in the stratosphere. He also found that CFCs were declining as they entered the stratosphere, thus supporting the theoretical expectations of Rowland and Molina. The reasons for his reservation about their theory was that he assumed other sources did account for chlorine loading in the atmosphere (Lovelock 1974).

In 1976, the first measurements of active chlorine (ClO) in the stratosphere were made. It also became clear that the substances were not washed out of the atmosphere by rain (*Science News*, 9 August 1975: 84). A sort of tacit agreement developed between the two camps that the presence of ClO in the stratosphere would provide the evidence that the Molina–Rowland hypothesis was correct. Some of the measured values for ClO concentration, however, lay far outside the anticipated range, a fact that industry exploited quite well. It is open to speculation whether industry would not have given up without a fight if the measurements and the model predictions had matched perfectly.

After measurements had shown that reactive chlorine was present in the stratosphere (if not in the concentrations expected, so that the question of the existence of sinks could not be conclusively answered), the next battle took place over the reaction rates. As indicated, industry would have been off the hook in the case that ozone depletion was occurring rather slowly. The first chemical species that slowed the depletion rate of ozone was chlorine nitrate ($ClNO_3$). This can form under certain circumstances from NO_x and ClO, both of which are themselves ozone destroyers. Ironically, it was Molina and Rowland themselves who delivered this resource into the hands of their opponents by drawing attention to it in an article. Industry could now once again argue that the problem had been relativised or was about to disappear. Modellers, however, determined that the predicted ozone depletion would still fall in the range of Molina and Rowland's original calculation of between 7 and 14 per cent, if at the lower end (Roan 1989: 76–8).

The life-span of CFCs and the closely connected question of tropospheric or stratospheric sinks became the topic of long and controversial discussion. As early as 1971 Lovelock set out to do measurements on a marine expedition from Wales to the South Pole. He did not get funding for this voyage from the British Research Council (see Chapter 2). As he told me, he was motivated by pure curiosity and paid for the voyage out of his own pocket. His results showed global concentrations of CFC 11 that tallied with cumulative world CFC production. He also found a north–south gradient of 7:4, that is, almost twice as high a concentration of CFCs in the Northern hemisphere. Since almost all sources of CFC emission lay in the more industrialised north, this result seemed to indicate a sink or a considerably shorter life-span than Molina and Rowland had assumed, although Lovelock himself did not make the link with residence times of CFCs. This was an issue taken up by Du Pont later on. Meanwhile, the CFC critics cast doubt on Lovelock's data. American researchers also made ship measurements that showed a much flatter gradient. The discrepancy could be explained by the fact that

Europe, from where Lovelock started his voyage, had a much denser pool of CFCs compared to the San Diego region from where the American ship set sail. At that time, Lovelock was blamed by some American scientists for inaccurate measurements. As a matter of fact, he was the only one who could measure CFCs down to concentrations of 40ppt.[3]

For the later development of the controversy two things are important. First, the above-mentioned international research programme carried out by the MCA to analyse the persistence of CFCs in the atmosphere. As it was put in an invitation to other CFC producers:

> Fluorocarbons are intentionally or accidentally vented to the atmosphere worldwide at a rate approaching one billion pounds per year. These compounds may be either accumulating in the atmosphere or returning to the surface, land or sea, in the pure form or as decomposition products. Under any of these alternatives, it is prudent that we investigate any effects which [CFCs] may produce on plants or animals now or in the future.
>
> (Manufacturing Chemists Association 1975: 2)

Second, there were several public statements (such as a full-page advertisement in the *New York Times*, 30 June 1975) from the Du Pont company, containing the following: 'Should reputable evidence show that some fluorocarbons cause a health hazard through depletion of the ozone layer, we are prepared to stop production of the offending compounds.' This voluntary commitment was to play an important role in the revision of Du Pont's CFC policy ten years later. At first, however, the controversy developed its own particular dynamic, in which it repeatedly fell to Du Pont to prove that there was no 'reputable evidence'. In the same announcement, the script for the next ten years was to be formulated: 'Claim meets counterclaim. Assumptions are challenged on both sides. And nothing is settled.'

Regarding the residence time of CFCs in the atmosphere, Du Pont proceeded until 1982 from the assumption of an extremely short life-span of halocarbons (10 to 20 years instead of 40 to 150, as assumed by Molina and Rowland; see Jesson 1982). Today it may seem that this was a trivial matter; supporters of regulations, especially, seem to think that matters were settled early on. In 1978, however, measurements of CFCs varied to a considerable degree, thus making calculations of their residence time extremely difficult. A report by the National Bureau of Standards, prepared for NASA, described the results of a collaborative test to evaluate the state of the art in measurement of the concentrations of halocarbons and nitrous oxide in the upper atmosphere. It states that '[a] pair of test samples, differing only slightly in concentration was sent to each of sixteen laboratories. Statistical analysis of the results reported by each laboratory indicates systematic differences between laboratories which are significantly larger than within laboratory precision' (Hughes *et al.* 1978).[4]

Research into the question of the atmospheric lifetime of CFCs began in 1978 with the Atmospheric Lifetime Experiment starting with the Adrigole station where Lovelock had set up and calibrated pioneering instruments. A network of measuring stations was then installed in Barbados, Samoa, Tasmania and Oregon. The finding, which was not published until 1983, was that the life-span of CFC 11 is approximately 80 years (Cunnold *et al.* 1983). Up to that point industry repeatedly attempted to use the alleged brief life-span of the substances as a toehold.

The question of the connection between a shrinking ozone layer, increased UV radiation and higher incidence of skin cancer was likewise taken up early, and was to become a long-running focus of the controversy. In 1975 Du Pont – with reference to one of the country's scientific experts, Frederick Urbach (Temple University) – attempted to give the impression that there was no immediate connection between UV radiation and the frequency of cancer. It seems that in this case use was made of sloppy transcriptions of a hearing in New York State that Urbach was given no opportunity to correct. When he heard of Du Pont's attempt, he made it clear that such a connection certainly existed: in his opinion a 1 per cent decrease in the ozone layer led to an increase of skin cancer by 2 per cent.[5] Urbach had made a similar statement before the Rogers Committee (see next section).

Institutional opportunity structures

At the institutional level there was initially chaos in terms of jurisdiction. For assessment of the scientific side two different institutions were involved (the NAS and the Climatic Impact Assessment Program, CIAP), while for the administrative aspects there were three (the EPA, the Food and Drug Administration [FDA] and the CPSC). Any type of political solution first required a clear definition of jurisdictions. These were both overlapping and incomplete. In the judgement of the Department of Justice, the EPA could regulate pesticides in spray-cans; the FDA could regulate them if they were contained in food, drugs or cosmetics; and the CPSC could regulate all other products in spray-cans, with the addition of refrigerators in private households and in schools, as well as air-conditioning units. No agency, however, could regulate the many industrial and commercial applications (for example, in automobile air-conditioners). For this reason the environmentalist group NRDC tried to settle matters by petitioning the CPSC. The CPSC was chosen because it was considered to have the most authority in terms of regulation. The NRDC demanded that the CPSC ban CFC use in aerosols. The CPSC declined by a vote of three to two, with reference to the pending NAS study.

In Spring 1975 Congressmen Rogers and Esch attempted to clarify the jurisdiction issue by means of a bill, in which the EPA was to be empowered to forbid the production and sale of CFCs, should their dangerousness be confirmed. Rogers took over the chair of the Subcommittee on Public Health and Environment, which held its first hearings on 11 and 12 December 1974 in the House of Representatives. Industry advocated a timetable of three years for additional research before anything could be said regarding the necessity of regulations, with the support of the atmospheric scientist McElroy.[6] Rowland and Cicerone were in favour of an immediate ban on propellants containing CFCs.

In January 1975 the National Science Foundation (NSF) and the Council on Environmental Quality (CEQ) resolved to collaborate in this matter. They formed the Ad hoc Interagency Task Force on Inadvertent Modification of the Stratosphere (IMOS), a committee to which fourteen agencies belonged. The goal of this task force was to assess the possible effects on the stratosphere and develop a plan of action. Mindful of the problem of jurisdiction, IMOS came to the conclusion that no CFC-specific law was to be passed. Instead, a regulation within the framework of the soon-to-be-passed Toxic Substances Control Act (TSCA) was envisaged.

IMOS held a hearing in February 1975, inviting Rowland, Cicerone, McElroy, representatives from Du Pont and other firms, as well as NRDC members (Bastian 1982: 171). Over a hundred observers took part in the meeting, among them representatives of the Canadian and British governments. A confrontation took place between Rowland and McElroy. An industry publication characterised McElroy as 'swimming hard against the tide of scientific experts that argue strongly that some sort of regulatory action should be taken early' (*Drug and Cosmetic Industry*, April 1975). As noted earlier, McElroy diverted attention onto methyl bromide – a move that was criticised by many as a diversionary tactic and an attempt to grandstand. Interestingly, the periodical *Drug and Cosmetic Industry* portrayed both scientists as coalition members of different interests: at the hearings, Rowland and Cicerone were supported by environmentalist Karim Ahmed (NRDC), while McElroy was backed by the Council on Atmospheric Sciences (COAS), the industry delegation.

IMOS published its study on 12 June 1975 and stated: 'There seems legitimate cause for serious concern.' The atmospheric concentrations of CFCs were judged to be consistent with total world production; the various rates of depletion, calculated by means of different models, were seen as being basically in agreement. The validity of these predictions could only be challenged if large unexpected sinks or major natural sources were discovered, which according to the current understanding was out of the question (IMOS 1975: 2). But opinions were split regarding policy implications. Several of the atmospheric scientists pushed for action as soon as possible, whereas the bureaucrats were rather more circumspect. A number of IMOS members saw no need for further research, since no fundamentally new developments were to be expected. They recommended beginning regulation immediately. Another group wanted to gain time, in the hope that the problem would disappear again or the Molina–Rowland hypothesis would be proven wrong.

A compromise was agreed that stressed the seriousness of the problem and regarded regulations as very likely sooner or later. It was decided, however, to wait for the report of the NAS. The possible deadline for regulations was set as 1978, in order to have further time for research and to grant industry the necessary time to adapt. Industry attempted in vain to delay or suppress the IMOS report at the last moment (Bastian 1982). When the report was finally published, the reactions were consistent with the political goals of the two alliances: industry saw it as an over-hasty anticipation of research findings; the environmentalists (NRDC) judged it to be to be too cautious (*Wall Street Journal*, 13 June 1975: 4). Du Pont tried to make the best of it in a press statement. The report was interpreted as saying that there was no immediate danger to the ozone layer and the health of the population and CFCs could be employed without hesitation until a conclusive clarification of the question (Du Pont, *Corporate News*, 12 June 1975).[7]

The report of the National Academy of Sciences was published in September 1976. Like the IMOS report, this study too was interpreted by both sides to their own advantage. Thus, it is hardly surprising that two major daily newspapers also offered contradictory interpretations the day after its publication. The *New York Times* of 14 September bore the headline 'Scientists support reducing propellants to protect the ozone layer', while the *Washington Post* of the same day read 'Science committee against propellant ban'. How could this happen?

First, it looked as if the NAS panel would confirm the Molina–Rowland hypothesis and declare itself in favour of regulating CFCs. However, new information about the chlorine nitrate reaction (see p. 109) brought utter confusion to the NAS committee. A fundamental reassessment of the scientific aspects and of the policy recommendations had to be carried out. The final published report had to contend with the problem that the ozone depletion rate originally planned for the report was halved from 14 to 7 per cent – which for the panel members constituted no justification for an unqualified advocacy of regulations.[8] The report was divided into two sections. The first was drawn up by the Panel on Atmospheric Chemistry and came to the conclusion that the Molina–Rowland hypothesis was essentially correct. The second was written by the Committee on Impacts of Stratospheric Change and recommended that the government wait another two years before resorting to regulations (Dotto and Schiff 1978: Chapter 11; Roan 1989: 79f.). The oracle had spoken, but what had it said?

> A one- or two-year delay in actual implementation of a ban or regulation would not be unreasonable. . . . As soon as the inadequacies in the bases of present calculations are significantly reduced, for which no more than two years need to be allowed, and provided that ultimate ozone reductions of more than a few per cent then remain a major possibility, we recommend undertaking selective regulation of the uses and releases of [CFCs].
>
> (NAS, cit. in Bastian 1982: 188)

Despite the ambiguous message, the confirmation of the Molina–Rowland hypothesis by the first report of the National Academy of Sciences represented an important resource for the advocacy coalition, particularly because the media interpreted the NAS report on the whole as confirming it (Dotto and Schiff 1978: 278).

If we go from publication, the summer of 1974, for a period of a little more than two years, we had this feeling that we were all out there alone. Until some seemingly independent organisation had said that 'Actually, this looks as if it makes sense', we were out all by ourselves. By we I mean Molina, Cicerone and myself.

(Personal communication from Rowland)

This was all the more significant because the NAS had the reputation of being a rather conservative committee, none too quick to push the interests of industry aside (Boffey 1975).

The media

The tremendous amount of media attention in the first three years after publication of the Molina–Rowland hypothesis is striking. Any subject with so much public attention almost automatically gets onto the political agenda (Kingdon 1984). This is due, first, to the spectacular nature of the subject. It seemed to belong to the realm of science fiction rather than that of science. It was hardly conceivable that the use of deodorant sprays could undermine the conditions necessary for life on earth. Might life on earth end not with a loud bang, but rather with a soft hiss?[9]

Second, influential and reputable journalists placed themselves on the side of the pro-regulatory coalition. Most prominent among these was Walter Sullivan, who published an ongoing series of reports in the *New York Times*.[10] The exposure gained through him, and the resulting focus given the subject by the most important American daily newspaper, drew the necessary public attention.[11] In the *New York Times* alone, the topic appeared on the front page eight times from 1974 to 1977. In the first decade of the controversy the *New York Times* published over a hundred articles, 67 of them in the first two years. The *Washington Post* and the *Wall Street Journal*, with no more than ten and 28 articles respectively, lag far behind.

Success of the advocacy coalition

The NAS report was published on 13 September 1976. Two weeks later IMOS met and recommended that the authorities take immediate measures. Shortly thereafter the head of the EPA announced that the 'non-essential' use of CFCs as a propellant was to be forbidden.[12] The FDA acceded to the NRDC's petition and announced that the three regulatory agencies would now proceed with the process of regulation. Representatives of industry expressed their surprise at this; in their opinion this ran counter to the NAS recommendation to wait for further research findings. Government representatives, however, agreed that this was not the case.[13]

The counter-alliance's hopes of gaining more time were in vain. IMOS had committed itself to the findings of the NAS report and the latter confirmed the plausibility of the theory, even if further research was considered necessary. In a way this was the compromise formula offered to industry: the set period of two years allowed for additional research. It was only that this time, unlike with IMOS, there was no higher authority to appeal to. A decision had to be made and regulations were advocated.

Predictably, the ambiguity of the formula gave rise to new controversies that were played out publicly and within the administration. The working group that was to hammer out the regulations consisted of representatives from the EPA, the CPSC, the FDA, the CEQ, NASA, the NSF, the Department of Commerce (DOC) and the Department of Transportation (DOT). The time frame within which regulations were to be passed was interpreted by one side to mean that the process should be completed within two years, while the other side interpreted it to mean that they should wait two years. The question was decided by the working group on the basis of the testimony given by the chair of the NAS committee (John W. Tukey) before the Bumpers Committee (Congress Hearings, 15 December 1976). Tukey said that he would not feel uncomfortable with the regulations already worked towards and that he could say the same of the other panel members. This was taken as strong circumstantial evidence that the NAS report recommended measures sooner rather than later (Wirth *et al.* 1982: 228).

NASA, which in 1975 had been charged by Congress to deliver an assessment by 1977, came to the view, in agreement with the NAS report of 1977, that newly measured reaction rates signified an exacerbation of the problem. Ozone depletion of up to 16.5 per cent was calculated based on 1975 emissions (NASA 1977: 6). During the period of public hearings reserved for appeals, industry tried to play one last trump. A new statistical technique allowed the ozone trend to be measured within five years. Regulations should be held off until an actual decline in the ozone trend could be established. The working group rejected this petition with the justification that should action prove to be necessary, the American regulatory system did not permit speedy action.

The EPA, under the amended version of the Clean Air Act (7 August 1977), was given the task of observing the repercussions of human action on the stratosphere by allocating and supervising appropriate research contracts, and reporting to Congress. It was also to report on the progress of international cooperation in protecting the stratosphere and make recommendations for further research (Wirth *et al.* 1982: 233). The battle between the precautionary principle and the 'wait and see' principle was decided by the Clean Air Act in favour of the precautionary principle. The Act empowered the head of the EPA

> to regulate any substance . . . which in his judgement may reasonably be anticipated to affect the stratosphere, especially ozone in the stratosphere, if such effect may reasonably be anticipated to endanger public health or welfare.
>
> (Clean Air Act, 42 USC §7457(b), cit. in Benedick 1991: 23)

Of central importance was the statement that the EPA required 'no conclusive proof . . . but a *reasonable expectation*'[14] was sufficient to justify taking action. This approach was to mark the American position in the subsequent international negotiations, too.

Third party support

The advocacy alliance also mobilised unexpected actors and resources, as the following two examples show. In the first case, shifting the burden of proof led to the mobilisation of new allies; in the second, additional allies were won because the problem was considered to be more serious than originally assumed. The analogy of criminal proceedings played an important role in this controversy from the beginning. A representative of the Du Pont company had argued that no one wanted a lynching party but rather a fair trial for the accused chemicals. The chair of the Council on Environmental Quality (CEQ), Russell Peterson, then coined the phrase: 'Chemicals are not innocent until proven guilty' (*New York Times*, 18 September 1976). This new framing of the problem, an inversion of the principle of burden of proof dominant in criminal proceedings, cleverly countered the argument of the chemical industry. Peterson's phrase carried weight, in particular because he himself had worked at Du Pont for over 25 years before he became Governor of Delaware and finally chair of the CEQ.

During the congressional hearings one congressman acknowledged that the CFC critics had formulated a serious problem responsibly, so that the perception of this case did not fit the standard pattern of the debate between environmental activists and industry:

> Mr. Chairman, when I came over here I thought we would have a hearing that would be a standard conflict between the environmentalists and industry and I can only say I am stunned by what we have heard here. There has not been inflammatory rhetoric or alarmist language but here we have some of the most distinguished scientists in America telling us about the problem.
>
> (cit. in Bastian 1982: 169)

> The responsible CEQ staffer, Dr. Warren Muir, Senior Staff Member for Environmental Health, later revealed that he had relished the assignment of sorting out the fluorocarbon question as an opportunity to earn a 'white hat' from industry by debunking the fluorocarbon theory, since he already had a 'black hat' because of his position on issues such as PCBs, asbestos, lead and toxic substances legislation. He soon came to realize, however, that the fluorocarbon problem was not going to disappear.
>
> (Bastian 1982, 169–70)

An additional deviation from the standard pattern of environmentalists versus industry was the fact that except for the NRDC, there was no environmental group involved in the case to any noteworthy extent. Above all, the NRDC followed a professional–legal strategy rather than a mobilisation based on panic-mongering.

An unexpected mobilisation of resources and actors is shown in the second example. After the discovery of the role of chlorine nitrate, which made the quantitative ozone depletion appear somewhat less dramatic, part of the advocacy alliance stressed that it could indeed be true that there was no great net loss of ozone (and therefore no increased risk of cancer due to more intense UV-B radiation), but that the result could be a change in the temperature profile of the stratosphere, and consequently have climatic repercussions. Although this frame switch was criticised as an evasive manoeuvre (in particular by McElroy, see Dotto and Schiff 1978: 257), it was nonetheless justified in the eyes of some scientists. IMOS formed a subcommittee on Biological and Climatic Effects Research (BACER) that was to investigate such climatic effects in a long-term programme (Bastian 1982: 181). In the first IMOS report these effects were not considered.[15]

If the chlorine nitrate problem and the frame switch in the direction of climate created short-term problems for the stability of the pro-regulatory alliance, they aided in the mobilisation of one new ally. Paul Crutzen prophesied that this branch of industry would soon be closed, if not on biological grounds, then for climatological reasons (*Science News*, cit. in Dotto and Schiff 1978: 257). The atmospheric temperature profile that could result from chemical reactions caused by CFC emissions appeared to him like that 'of another planet'. Crutzen's statement can be interpreted as that of a well-disposed critic (see Figure 1.3, p. 25) who considered the original data of the MRH to be even more significant than the authors themselves had initially seen. Now the thesis that CFCs were in and of themselves already an efficient greenhouse gas was taken up again. As one of the early advocacy scientists put it,

> In 1975, Ramanathan came out with the first calculation that CFCs would contribute to the greenhouse effect. R., [another scientist] then at General Motors, ran her model later in 1975, and basically confirmed Ramanathan's conclusions. But, she was not allowed to publish this result because (a) Ramanathan was then not very well known, and his paper was being questioned – therefore not yet accepted, and (b) knowledge that CFCs had a greenhouse effect might be of competitive use to General Motors versus Ford or Volkswagen. I think also that confirmation by General Motors would have been taken as industrial acknowledgement of the situation, foreclosing on the option to ignore Ramanathan, or to throw doubt on his conclusions. . . . R. was not only not allowed to publish – her copies of her work were confiscated by General Motors. She did think that it was important, and told me. In Geneva, later that year, when Ramanathan's work came up, and one of the meteorological types questioned it, I was able to say that the work seemed to be quite credible, and in fact had been confirmed by one industrial scientist not at liberty to publish results. There was no publication of the confirmation, but the grapevine knew that Ramanathan was basically correct.[16]

Rhetorical strategies

As has already been made clear, both camps saw the controversy's preliminary decision in the judgement of the two evaluating institutions, IMOS and the NAS. They therefore attempted to influence these agencies' judgement during the hearing process by any means available. And even after the reports were concluded, they tried to appropriate the verdict in their own favour.

The CFC critics perceived the recommendation of the NAS to reserve a further two years for research before proceeding with the regulation as interference in political affairs. David Pittle (CPSC), who sympathised with the advocacy coalition but remained in the minority in the 3:2 vote on the NRDC petition, saw in the decision not a scientific problem, but rather a social question, in which many partially conflicting considerations played a role (*Chemical and Engineering News*, 27 September 1976). He criticised the NAS for rendering a *political* assessment with its statement 'two years' wait can't do any harm'. Pittle limited the committee's function and the value of its statement to the technical aspects; policy recommendations were seen as falling outside its remit. The NAS policy recommendations should carry no more weight than the recommendations of other informed citizens (*New York Times*, 18 September 1976).

A Du Pont spokesman pursued just the opposite strategy. He stressed the importance of the most prestigious scientific institution in the country, which had come to the conclusion that a further two years of passivity could do no great damage. He tried to exploit the reputation of a scientific institution for political purposes while simultaneously bracketing out the result that the same institution considers regulations to be unavoidable. The hope was that by continually gaining time, one might still nonetheless find a way to refute the theory or at least to eliminate the pressure caused by the problem.

In view of the uncertainties in the model calculations, there was indeed no purely scientific method to decide whether CFCs should be regulated, and if so, how strictly. Here each side had its own reading. Industry followed the slogan 'innocent until proven guilty', while the critics deduced the need for extraordinary precautionary measures. The question was whether the uncertainties of the computer models represented an argument for or against regulation. An error factor of 2 in the models meant that the problem could be either half as big or, just as easily, twice as big, as the atmospheric scientist Ralph Cicerone stressed at the hearings (United States Senate, Congress Hearings 1975: 949, statement Cicerone). Uncertainty cuts both ways.

Meanwhile, many states had embraced regulations of their own. Ten states joined the petition of the NRDC. Oregon and New York passed measures to ban spray-cans containing CFCs in future, and 13 other states had similar laws in preparation. In such a situation it is convenient for the CFC producers above all to have *uniform* regulations, even if these lead to the restriction of a range of applications. In addition, one section of industry broke ranks. Although it was neither a producer nor a major user of CFCs, the Johnson Wax Company dealt the CFC alliance a psychological blow in June 1975, when it announced it would no longer employ CFCs in its products.

The reaction of consumers was clear. Between 1975 and 1977 the use of CFCs in private households declined by over 20 per cent. As early as 1975, an opinion survey carried out by the chemical industry clearly showed a slump in the popularity of spray-cans (*Aerosol Age*, June 1975).

Practical judgement

The three advocacy scientists who formed the nucleus of the pro-regulatory alliance developed notable skills in their dealings with the public and the mass media. In the course of innumerable events involving the lay public and hearings before Congress they succeeded in portraying the dangers convincingly. They developed a sensible political strategy for the protection of the ozone layer that was far removed from fundamentalist proposals. Instead, they made suggestions towards a technical solution to the problem. Thus they were able to distance the controversy from the 'industry versus environmentalists' model. First, they proposed substitutes for non-essential applications of CFCs, particularly as a propellant in spray-cans. For the USA that would already produce a reduction in emissions of about 50 per cent. This pragmatic proposal was based on the calculation that substitute products were available (for example, pump sprays, propane and butane mixtures as propellants) or could quickly be developed; that this was the greatest single non-essential application; and finally that this use constituted a symbol of frivolous luxury consumption.[17] Other applications, for example, in refrigerators and air-conditioners, would be spared for the time being. This calculation was so successful that it found its way into the regulations. Moreover, this group argued, in making this decision under uncertainty, the potential benefit for the planet, should new findings prove the danger to be less than had been assumed, was incomparably greater than any potential harm.[18] In the mid-1970s, more than half of the annual world production of approximately 700,000 tonnes of CFC 11 and CFC 12 went into aerosol propellants. In the USA, most of this was used in personal hygiene sprays.

The initial definition of the problem and proposals for a possible solution made by the pro-regulatory alliance had a structuring effect for the whole controversy. An atmospheric scientist characterised Molina and Rowland's early work as follows: 'With it they'd already cleared the field, they didn't leave out a single aspect that could have been thought of at the time' (GEAS 26; this comment presumably refers to their 35-page paper in *Reviews of Geophysics and Space Physics*, not to the original two-page publication in *Nature*). A relatively small nucleus of actors managed to mobilise resources and new allies in a relatively short time.

In summary, it can be said that various processes mutually reinforced one another. The media attention led to the topic being placed on the political agenda. The Congress Hearings created an additional public. The committee chairs were favourable to the critical scientists. Individual states began to pass laws, and consumers turned away from spray-cans. In order to legitimate a federal regulation, the ministries gathered in IMOS awaited the vote of the NAS. Thus the NAS was granted a crucial role; in a way it decided the outcome of the policy dispute.

A contrasting example: the case of Germany

Political context

It is probably no exaggeration to say that the problem of environmental protection at the end of the 1960s was imported from the USA.[19] True enough, the SPD (Social Democratic Party) had fought its electoral campaign as early as 1961 under the slogan 'Blue Sky over the Ruhr'; and even earlier, an inter-parliamentary working group had laid out principles for environmental protection (cf. Burhenne and Kehrhahn 1981). However, these were sporadic attempts that failed to place the topic on the long-term political agenda. The impetus to politicise environmental protection came in 1970, at the behest of international organisations, the European Council, the United Nations, NATO and the OECD. In 1972 the first report of the 'Club of Rome' appeared (Meadows and Meadows 1972), and in the same year the first UN conference on the environment took place in Stockholm for which the government of the Federal Republic also prepared its agenda.

While the concern with the environment came onto the West German agenda from outside via international organisations, American approaches were studied closely within the parties and among Ministry officials. In 1969, in the course of administrative reforms, environmental protection was placed under the jurisdiction of the Ministry of the Interior. Excepted from this transfer were landscape and nature conservation, which remained under the authority of the Ministry of Agriculture. The Federal Ministry of the Interior was a department held by the Free Democratic Party (FDP), in which the Secretary of the Interior[20] Genscher (later known for his role as foreign secretary) attempted to realise a policy of 'out with the old, in with the new' by means of the Emergency Programme for Environmental Protection, passed in September 1970. Genscher was supported by Social Democrat Ehmke in the Chancellor's office, who was himself much impressed by American environmental activities. On the initiative of Chancellor Brandt, several working groups and committees were formed to deal with environmental issues. At the end of 1970 the programme-planning department for 'environmental coordination' was established in the Ministry of the Interior. Its head, Peter Mencke-Glückert, had likewise been enthused during a stay in the USA by American environmental policy. His formulation of the programme was heavily inspired by American examples.

> He is a man with a great capacity for enthusiasm, with many good ideas, a thorough knowledge of the international research and considerable ability to assert himself. The draft that he quickly produced was a complete success, and in fact it became a blueprint for European environmental policy. It was far in advance of the development of the public consciousness.
>
> (Genscher 1995: 127)

The environmental programme passed in 1971 by the federal cabinet served as the example for a multitude of similar programmes established by German parties and associations. In this programme were formulated the three basic principles of German environmental policy that have remained valid to the present: the principle of precaution, the 'polluter pays' principle, and the principle of cooperation (Müller 1989). Although it was legally mandated, the precautionary principle played no prominent role in the discussion of CFCs in the 1970s (cf. Weidner 1989). This is in marked contrast to some claims made in the recent literature in environmental sociology that Germany is in the forefront of ecological modernisation because it has allegedly implemented the precautionary principle (for example, Dryzek 1997).

In the same year, the Ministry of the Interior established the Council on Environmental Questions (Sachverständigenrat für Umweltfragen, SRU), which was meant to take on an important function in support of environmental protection: by analogy to the American example, it was to issue periodic reports pointing out setbacks and proposing measures.[21] If, on the one hand, the legal status of the SRU corresponds more to that of an auxiliary agency of a Ministry than to that of an independent expert group, it is, on the other, 'in terms of actual form more like the economic council than a ministerial advisory board'; in other words, its institutional role is rather weak (Timm 1989: 114). The SRU's main problem is that it is caught in the dilemma of weighing contradictory goals against each other, namely to reconcile economic and ecological viewpoints. Thus it does not seek to pursue environmental policy as its main priority. According to the SRU this is done only in cases where a fundamental deterioration of living conditions is imminent, or when the long-term security of the population's basic living conditions is endangered (Rat von Sachverständigen für Umweltfragen 1978: 577; cf. Timm 1989: 239). The orientation towards balance makes it clear that the SRU does not take the position of a speaker for diffuse environmental interests. At the very beginning of the environmental report for 1978, the board's objectives are described thus: 'Its reports should . . . not be one-sided or biased pleas for environmental protection' (Rat von Sachverständigen für Umweltfragen 1978: 13). The Federal Office for the Environment (Umweltbundsamt, UBA), set up in Berlin in 1974 as an advisory authority of the Ministry, remains in separate quarters from the SRU. The UBA and SRU hardly have the resources to conduct their own research (H.-J. Luhmann 1991).

As in the USA, it was realised that the systemic character of environmental problems demands a fundamental reorganisation of the federal government's management structures. This was the central message of the expert committee chaired by Georg Picht. Similar to Nixon's guidelines for the EPA, it was determined that the existing division of responsibilities led to a departmentalisation of the problems (into the environmental media water, air and soil), thus making their solution impossible.[22] On the one hand, the Picht Commission did support the Ministry's efforts to acquire researchers and consultants of its own; on the other hand, however, it came out against the establishment of new federal or regional institutions. The monodisciplinary expert knowledge produced in the existing institutions was judged to be adequate, insofar as it could be summarised by a team of advisers mediating between scientific information and political action.

However, the government did not follow the Picht Commission's proposal, but rather decided in favour of a solution that excluded the systemic-ecological aspect. The responsibilities would be divided up among existing departments (Küppers *et al.* 1978). What is more, the UBA, subordinate to the BMI, was established without the recommended scientific institute. Although the environmental programme of 1971 still expressed the intention of setting up a Federal Office for Environmental Protection (with a staff complement of 850 workers), the number allotted at the establishment in 1974 of the UBA, which was much more restricted in its function, was no more than 173, rising by 1982 to 439 (Müller 1986: 70). In hindsight, Genscher drew a sobering balance sheet:

> On 23 July 1974 the law regarding the Federal Office for the Environment, the UBA Law, came into effect. Many points of the original design, in the course of a laborious process of co-ordination with numerous affected departments, had fallen by the wayside. As a result, probably the oldest and best-known environmental institute in Europe, the Institute for Water, Soil and Air Hygiene of the Federal Health Office, although its seat was in Berlin, could not be incorporated into the new office.
>
> (Genscher 1995: 134)

Not until 1994 was the Institute for Water, Soil and Air Hygiene handed over from the Federal Health Office to the UBA (Fülgraff 1994). Now in the UBA as well, at least, great pains are taken to overcome the division of the environmental media water, air and soil.

Analysing the first decade of environmental policy in Germany, Müller (1986) differentiates three phases: an offensive phase (1969–74), a defensive (1974–78) and a recovery phase (1978–82). The actual beginning of the defensive phase falls in 1975, when the government discussed the reactions to the economic crisis in close consultation with industry, the unions and the environmental protection department. During this meeting, which became known as the 'Gymnich Talks', industry and unions formed an alliance that placed three theses on the agenda: the curtailment of investment caused by prolonged legal proceedings, the endangerment of the energy supply posed by too strict constraints on air pollution, and the threat to jobs posed by environmental protection. The government largely endorsed this alliance. For economic reasons cuts were to be made to central principles of the environmental policy. As a result, it would have been possible in exceptional cases to replace the 'polluter pays' principle with the principle of a common burden.

> Chancellor Schmidt said that environmental protection had to lower its sights because of the costs. This applied to the catalytic converter for cars, but also to all of the projects and ordinances in the field of chemistry. That was all

influenced by the coal and steel industry and the Ruhr area [North Rhine Westphalia]. It's very important for an SPD chancellor to hold the biggest SPD province. This included the ordinance against large fire combustion plants, which was put off until the beginning of the 1980s. There were economic arguments against it.

(GEAD 12)[23]

The only environmental problem that drew attention in the second half of the 1970s was the problem of nuclear power, which was so dominant that it over-shadowed all other problems (Müller 1986: 104). In the defensive phase, the government implemented some laws that had already been planned and pre-pared, but it did not introduce new initiatives; though, indeed, no environmental laws were repealed. The environmental policy of the Ministry of the Interior however, went on the defensive. Terrorism took up the attention of the Secretary, who was furthermore isolated by the growing strength of the economic wing within his party. The resulting strategy of conflict avoidance robbed the advocacy alliance of a potentially vociferous ally and a possible speaker. In the recovery phase (1978–82) the Secretary of the Interior regained the initiative, not least by means of the politicisation of environmental protection resulting from the elect-oral victories of the green parties, who had taken advantage of the anti-environmental turn of the government. When green groups formed themselves into a party, the young generation was particularly well represented. The Secre-tary's willingness to enter conflict, however, was constrained by his own bureau-cracy, which had grown so accustomed to the consultative mode that it persisted in a state of 'learned helplessness' (Müller 1986: 466). As a result, the newly arisen freedom of action could not be fully enjoyed. In 1979 the Permanent Secretary of the Ministry of the Interior announced that the time was ripe for an 'ecological turning point (Wende)', but was unable to push this through in the face of the Ministries of Agriculture and of Industry, who still retained partial responsibility for environmental affairs (Müller 1986: 128ff.). Until the end of the social–liberal coalition in 1982, the topic of the environment never became an issue during elections, not even in the federal elections of 1980.

Scientific context

The stratosphere was not a focus for research in Germany in the mid-1970s. It only gradually developed as a result of the CFC problem. Aerial chemistry had its mainstay in the troposphere and was pursued by many researchers of inter-national reputation, for example, Professor Junge of the Max Planck Institute for Chemistry in Mainz. From 1979 to 1985 the German Research Association (Deutsche Forschungsgemeinschaft, DFG) financed the special research area of 'Atmospheric Trace Substances' at the Universities of Mainz and Frankfurt, led by Georgii, Jaenicke, Junge and Warneck (DFG 1985: 30ff.).

News from the United States

American developments made their way to Germany through two channels.[24] One was the German Embassy in Washington, which periodically passed its assessments to the German government via the Foreign Ministry (Auswärtiges Amt, AA) and the Ministry of the Interior (BMI). The other was as a result of personal contact between Rowland and an employee of the UBA who had likewise become environmentally interested during a stay in the USA.

Policy makers in Germany looked to the American discussion and wondered: what does that mean for us? In view of the American superiority in this field, it was decided that Germany could make only a modest contribution in terms of research. In an internal note in November 1974, the Ministry of the Interior deemed two pieces of information as important: the level of CFC production in Germany and the degree of air pollution caused by these substances. Contact was established with Hoechst AG, one of two CFC producers in Germany (the other being Kali-Chemie). The two firms together produced approximately 15 per cent of world production. The political motto in regard to regulations was wait and see what the research in the USA produces. Conclusive findings were not expected before 1977.

The German Embassy in Washington communicated to the AA in July 1975 that the IMOS study would have no immediate effects, since the Americans were waiting for the report of the US National Academy of Sciences. In August 1975 the Research Ministry (BMFT) answered a question in the Bundestag to the effect that the government did not consider the Molina–Rowland hypothesis to be sufficiently proven as to draw any consequences. The government wanted to wait for the findings of research projects underway in the USA in order to 'be able to carry out a conclusive assessment of the problem'. Thus the shape of German policy was to a great extent fixed; the argument that not enough was yet known to be able to draw conclusions was to be repeated just as often in subsequent years as the hope for a future conclusive assessment of the problem.

In December 1975 the German Embassy informed the government that a bill for the regulation of CFCs had been placed before the US Congress and that its passage was considered likely. In this bill, the EPA was charged with the coordination of research in the area of CFCs and ozone, which was to be implemented in regulations after, at most, two years.

German reactions

At the same time, scientific cooperation was arranged between the Max Planck Institute for Chemistry in Mainz and Hoechst AG. The MPI scientists stressed the necessity of gathering precise data on world production of CFCs, which the Hoechst representatives promised – though it was to be more than ten years before this promise would be fulfilled and the German production statistics were released. In late September 1975 a meeting took place with a view to a German research programme, at which scientists, representatives from Hoechst and the

UBA participated. The following foci of future research, among others, were listed: sources and sinks for CFCs; instrumentation; measurements in the stratosphere; model-based calculations; and effects of UV radiation.

The first point was seen as a potential exoneration of industry and hence industry was to bear a share of the financing. In 1976 the research programme began; 14 planned projects with an average duration of three years and a total budget of 14.7 million DM were financed.

Administrative responsibilities were accorded to the Ministry of Research and Technology for scientific aspects, to the Ministry of the Interior for data on production and consumption, and to the UBA for the coordination of all R&D efforts. Moreover, the UBA was to work out the bases for potential control measures. The German position for a planned international meeting in Washington in 1976 appeared as follows: the scientists' fears of possible ozone depletion were being taken seriously, but for the moment the available findings permitted no assessment.

Policy networks

Using the approach developed in Chapter 1 and applied above, in this case, too, we can again differentiate between the most important actors, resources and new allies. Although only in embryonic form, in Germany there were also two alliances, each holding contrary positions, making claims and framing the subject accordingly. The UBA and Hoechst could be described as mouthpieces of their respective 'senior partners' in the USA. Conspicuously lacking in Germany is any equivalent for the US IMOS Task Force and the US National Academy of Sciences. Also missing in Germany during the first years of the controversy were public parliamentary hearings on this question. The pro-regulatory alliance, moreover, was hampered by the fact that the UBA was subordinate to the Ministry of the Interior, showing the restraint of a newly established agency and taking no line of its own in public. In the first two years of the controversy there was no independent stratospheric research in Germany. All data originated in the USA and in Great Britain. The debate between the pro-regulatory alliance and its opponents bears all the features of a proxy war.

In Spring 1975 the UBA established contact with Dieter Ehhalt and Paul Crutzen (NCAR and NOAA, Boulder, USA). In 1974 Ehhalt returned from the USA to the nuclear research facility in Jülich. He had been a member of the NAS panel in 1974–75. Crutzen considered production caps in the USA to be very likely and saw a high level of correspondence between the amounts of CFC 11 produced and concentrations measured. This meant that tropospheric sinks could be ruled out, and that the substances must have a rather long life-span.

Rowland visited the UBA in September 1975 and again in March 1976. During the first visit he declared himself ready to collaborate in the presentation and interpretation of the scientific knowledge. He also counselled the UBA, however, to accept a description from industry's point of view as well, in order to forestall the criticism of a one-sided interpretation. At his second visit the question of tropospheric sinks, which he held to be non-existent, was the focus. He referred, moreover, to Anderson's planned balloon measurements, which were supposed to measure chlorine monoxide (ClO) in the stratosphere. Proof of ClO could be taken to verify the Molina–Rowland hypothesis.

In mid-May 1976 the Battelle Institute published a study ordered by the Ministry of the Interior. This study considered the Molina–Rowland hypothesis to be essentially correct, but held individual steps and conclusions to be insufficiently proven, in particular the effect of more intense UV radiation on plants. In summary, the role of the UBA as a critic of CFCs was precarious. It had no great desire to raise its profile, and showed consideration for the needs of industry and of its superior agency (the Ministry of the Interior) by treating the matter as confidential.

The opposing coalition

The mouthpiece of the counter-alliance was Hoechst AG. Hoechst, like the counter-alliance in the USA, also attempted to find 'holes' in the Molina–Rowland hypothesis. In early 1975 Hoechst estimated the total world production. In contrast to Rowland and Molina's claim, Hoechst saw no congruence between cumulative world production and atmospheric concentration. The figure given by Hoechst is considerably higher than all other estimates, leading to the suspicion that they were chosen in order to make the existence of sinks seem plausible. Hoechst's numbers were in the area of 10 million tonnes CFCs 11 and 12. Du Pont estimated the total world production to be approximately 6 million tonnes. Other parts of the Molina–Rowland hypothesis were also contested by Hoechst, such as the insolubility of CFCs in water and even the photolysis in the stratosphere (Report of the UBA pilot station in Frankfurt, 22 January 1975).

In October 1975 Hoechst went on the offensive. With reference to Lovelock, Hoechst claimed that at most 20 per cent of the chlorine content of the atmosphere could be attributed to CFCs. On the authority of Watson, the rate constants of two reactions significant for ozone depletion were corrected downward by a factor of 1.6 and 2 respectively. With reference to Crutzen, the extent of hypothetical ozone destruction was calculated to be two to three times smaller than originally assumed. It was further argued that stratospheric ozone quantities demonstrate a natural fluctuation much greater than the loss allegedly caused by chlorine. Hoechst posed the rhetorical question whether this problem 'still merits even an above-average degree of ecological attention'.

After the publication of the NAS study Hoechst could no longer hold to this line; NAS had recommended taking regulatory measures within two years. Hoechst now built its argument on an oceanic sink, which would decrease ozone depletion from 7 to 6 per cent. Furthermore, the relative contribution of various CFCs to ozone depletion was stressed; CFC 11 contributed only 2 per cent, whereas CFC 12 contributed 4 per cent. Hoechst could not quantify this effect, however, since this would require complicated modelling calculations that obviously could not have been performed by Hoechst itself (Letter to the UBA, 1 December 1976).

Two things are apparent here. First, Hoechst cast doubt on essential parts of the Molina–Rowland hypothesis that in the USA were widely accepted even by Du Pont; second, having no research of its own to make use of, Hoechst fell back on (over-) interpretations of other sources and in-house ad hoc assessments. Du Pont was more open to scientific research, as was expressed by the fact that, among other things, scientists of good reputation were engaged and efforts in the field of modelling were undertaken.[25] Hoechst was exclusively engaged in defending its product. A small but telling aside: Hoechst often uses the abbreviation FKW (FC) or the expression 'fluorocarbons' (*fluorierte Kohlenwasserstoffe*). Thus the actual critical component, chlorine, is suppressed.

In August 1976 representatives from Hoechst and Kali-Chemie discussed the problem in the Ministry of the Interior. The representatives of industry were of the opinion that damage to the ozone layer from CFCs was 'even less proven today' than in the previous year. They feared, however, that a ban on CFC use could nonetheless come to pass in the USA for political reasons. They indicated that the chlorine content of the atmosphere could only be raised by about 10 per cent due to propellants. This opinion went uncontested. The Ministry of the Interior shared the attitude of industry that there was no need for legislative measures. The legal basis for this view was § 35 of the Federal Emissions Protection Law, whose implementation requires that damage be proven (this was regarded as the legal basis for action, not the famous precautionary principle).

Hoechst informed chosen sectors of the public via the trade paper *Frigen Information* (Frigen was the trade name for CFCs produced by Hoechst). Due to the 'unobjective and speculative use' of information, a 'broad dissemination of information on the topic of ozone at the present moment' was considered inappropriate. The periodical *aerosol report* provided commentary on the controversy during this entire period. At the very beginning, the Molina–Rowland hypothesis was seen as doubtful on all points; particularly the dismissal of tropospheric sinks. Instead, calculations were made to show how hydrolysis of CFCs in the oceans could occur. This was assumed 'with a probability bordering on certainty'. It was conceded that 'until the availability of conclusive proof by means of measurement' one 'should rely on common sense in regard to aerosols'. In contrast to American models, which were obviously suspect because they were calculated by computer (can we still imagine a time before computers?), the defiant claim was made: 'For our calculations you don't need a computer. All you need is a slide-rule and a sheet of paper' (*aerosol report* (1)1975: 15).

In 1977 a scientist from Hoechst AG contributed to the periodical *Nachrichten aus Chemie und Technik*. He summarised the research findings of the NAS report, in order to further mention new scientific findings that 'might lead to considerable corrections'. Here three points were listed:

- the possibility of photolytic depletion in the troposphere is again asserted;

- with reference to a newly determined rate constant in Cambridge (UK), a reduction of the predicted depletion by 20 to 30 per cent is assumed;
- mention is made of the unexpected ClO and OH concentration profiles obtained by balloon measurements, as carried out by Anderson in particular, which contradict model predictions.

The studies of the American scientist Howard on the reaction of HO_2 and NO were cited, albeit without giving away that this reaction proceeds 40 to 60 times faster than previously assumed and thus drastically *exacerbates* the problem. The rest of the article, in terms of the state of the controversy, was less surprising:

> This brief enumeration of recent research findings shows that *many questions are still open* and thus that the scientific discussion of the CFC–ozone hypothesis is still very much in progress. The CFC–ozone hypothesis could probably only be confirmed if a balance of the most important chlorine compounds in the stratosphere is found and the concentration profiles . . . are in tolerable agreement with the models.
>
> (Russow 1977: 508, original emphasis)

The article closed by relativising the risk of cancer and emphasising the economic importance of CFCs, including the difficulty of shifting to alternative propellants.

Hoechst versus UBA

If we examine the proxy war between Hoechst and the UBA more closely, the following pattern of communication can be observed: Hoechst draws the attention of the Ministry of the Interior to the fact that new data or reports from the USA are available. These are, for the most part, selectively chosen data that defuse the Molina–Rowland hypothesis. They are provided to the Ministry, which then passes them on to the UBA with the request for comment, often stressing the urgency of the matter. Since the UBA does not have the resources at its disposal that would be required to do its work effectively, the superior agency often got the impression that Hoechst could provide good reasons for a 'wait and see' attitude.

Hoechst repeatedly released statements sceptical of the validity of certain research findings. In an assessment delivered to the UBA in June 1977, Hoechst repeated the thesis of a tropospheric sink, this time with reference to a researcher from the US National Bureau of Standards, according to whom long-wave UV light could already destroy the CFCs on the ground (cf. Russow 1977). The average life span of CFC 11 is given as 20 to 30 years, which would lead to an ozone depletion of 2.5 per cent as compared to 7.5 per cent predicted by the NAS. Hoechst was confident, moreover, that new research findings would lead to a further drastic lowering of this value. In addition, the repercussions of increased UV radiation, in particular the risk of cancer, were played down: 'More than 95 per cent of all skin cancer cases can be treated easily, efficiently and relatively inexpensively as out-patients.'

In September 1977 Hoechst once again turned to the UBA to make clear that it did not agree with a recent assessment made by Crutzen, who estimated 12 per cent ozone depletion. With this, Crutzen moved into the range of the estimate given by NASA (1977) of 10.8 to 16.5 per cent. Crutzen arrived at this finding on the basis of Howard's newly calculated rate constant (see above). Hoechst indicated that comparative measurements in Göttingen had not confirmed Howard's values and that therefore the 12 per cent value was not acceptable, 'not even as a basis for discussion', as long as the contradiction remained unresolved. (In the further course of the debate, nothing more is heard of this measurement in Göttingen.)

The UBA sought the support of scientists in order to be able to check the new findings cited. One scientist to whom they wrote never answered. Rowland, in the meantime, reported six successful balloon measurements carried out by Anderson (he fails to mention the fact that several measurements did not match the theoretical models, since they were unexpectedly high) and corrected the report on tropospheric sinks: the National Bureau of Standards researcher had himself made clear that this sink was significant not for CFCs, but rather at the most for CCl_4 (carbon tetrachloride). Rowland reported the state of knowledge a year after the publication of the NAS reports as follows:

- the chlorine chain has been confirmed;
- the north–south gradient is flat (<10 per cent difference, i.e. there is no sink);
- chlorine nitrate plays no significant role;
- the predicted ozone depletion rate climbs from 7 to 12 or even 15 per cent.

The UBA did not counter-attack with this information from Rowland, however, but rather came to a written agreement with Hoechst that, until the reaction rate in question was clarified, the values of NASA (and of Crutzen) would provisionally be accepted (Letter to Hoechst of 28 November 1977).

In Spring 1978 Hoechst declared, with reference to a study carried out at the behest of the MCA, that a statistical analysis of the available ozone measurement data (trend analysis) could determine no change in the ozone layer, although according to the Molina–Rowland hypothesis the layer should already have decreased. The UBA replied that these trend analyses were correct, but that the conclusion that the Molina–Rowland hypothesis was disproved did not follow. It was out of the question that a presumed annual ozone depletion of 0.1 per cent could be perceived in the high natural fluctuation of the ozone content. It regarded measurements at an altitude of 30 to 40 kilometres more promising, where the main depletion was indeed anticipated. To this end, hope was placed in appropriate measurements from NASA satellites. Hoechst yet again stressed that there were too many discrepancies between the model calculations and the values measured to consider the Molina–Rowland hypothesis proven. Hoechst was confident that even small ozone reductions could be measured by the WMO measuring network (down to 1.5 per cent per decade). The purpose of this claim is evident: if it is possible in principle to see such tiny changes, but in reality they cannot be measured, then this would be evidence of a constant ozone layer. The UBA, in its reply, appealed to NASA's statement that the threshold value measurable by the WMO measuring network was 5 to 6 per cent ozone depletion per decade.

Atmospheric scientist Ehhalt from Jülich summarised the scientific findings in late 1977 for *aerosol report*. He stressed the uncertainty of the rate constants and the problematic nature of Anderson's ClO measurements, which were in part eight times higher than anticipated. Because of the problems with the rate constants, he would consider the degree of uncertainty of the model predictions to be greater than the NAS had assumed. Presuming that Anderson's abnormally high ClO concentrations could not be attributed to measurement errors, then there must be – besides CFCs – other sources of stratospheric chlorine. In contrast to Hoechst, however, Ehhalt took seriously Howard's newly measured rate constant for $NO + HO_2 \rightarrow NO_2 + OH$. According to Ehhalt:

> On the basis of this rate constant the predicted O_3 reduction becomes about twice as large as the prediction of 7 per cent made by the NAS committee. This means that the danger of a reduction of the ozone layer is considerably greater than was assumed in 1976.
>
> *(aerosol report* (12)1977: 456)

He stressed that the research of the last two years had increased knowledge enormously, but that the imprecision of the models had at the same time become greater. In order to limit the margins of error in ozone prediction, much more research was needed. The tone of the article betrays some doubt regarding daring American plans, since the uncertainty factors had in fact been underestimated. This article, while very objective overall, can be interpreted in two ways. One, Ehhalt emphasises the uncertainties of the models; two, he draws attention to the possibility that the danger had been underestimated. Everything depends on an interpretation of these two factors – an interpretation that the author does not provide.

New allies

Industrial and consumer organisations took the attitudes that might be expected. While the Industrial Association clearly formulated the 'wait and see' principle, the Consumers' Association (Arbeitsgemeinschaft der Verbraucher, AgV) pursued a mere watered-down version of the precautionary principle. In 1975 the AgV demanded that the German government make a statement and, if necessary, place a ban on CFCs. Due to the government's reticence, the following year the AgV recommended limiting the purchase and consumption of spray-cans 'until the scientific investigations in progress worldwide have provided a definitive explanation'. The AgV reported that Permanent Secretary Hartkopf (Ministry of the Interior) and Federal Research Minister Matthöfer had received the call for limiting spray-can purchase positively as an interim solution (*Verbraucherpolitische*

Korrespondenz, 18 January 1977). The Association for the Personal Hygiene and Detergent Industry (Industrieverband Körperpflege and Waschmittel e.V.) expressed to the UBA its agreement that, in view of the unclear scientific situation, it would make no decisions with far-reaching consequences. In support of this position, a study by the Clean Air Council of Great Britain, which had stressed the scientific uncertainty and the necessity of further research, was cited. Great Britain, too, wanted to wait and see, and in the meantime was encouraging both that alternatives be produced and that leakage from cooling systems be minimised.

The SRU, in its environmental report for 1978, considered it

> appropriate that the producers and users of CFCs in the Federal Republic of Germany work out and present a plan by which the emissions can be clearly reduced. This applies in the first instance to the use of CFCs as propellant gases.
>
> (Rat von Sachverständigen 1978: 175)

This rather modest declaration confined itself to vague generalities.

Informal solution: maintenance of the status quo

Shortly after the publication of the NAS study in September 1976 and the completion of the Batelle report, the UBA invited the affected industrial interest groups and companies to discuss the necessary measures. The UBA proposed voluntary limits for industry and alternative propellants for spray-cans. There was hope that market mechanisms alone would achieve a sufficient restriction on CFCs so that state intervention could be avoided.

Industry representatives considered a voluntary lowering of CFC consumption by one-third relative to the base year 1975. However, in view of the expected pressure from the American bans, the UBA wanted to lower consumption by 50 per cent. In late 1977 a voluntary agreement was reached between the industry association VCI and the Ministry of the Interior involving a limit on CFC use in aerosol products. According to the agreement, CFC deployment was to be reduced 30 per cent by 1979 relative to the base year 1975.

The German government wanted to effect, as completely as possible, a voluntary end to the use of CFCs as aerosol propellants within a few years. This was to occur through pressuring the producers and informing the consumers. There was no thought of bans, since no one wanted to endanger jobs. 'If, however, developments come to a halt, CFCs can be banned at relatively short notice', as a statement of the Bundestag's Petitions Committee put it in June 1977. Permanent Secretary Baum was even clearer in his response to a parliamentary question. He referred to the ongoing discussions with industry and made it clear that a legal regulation was being worked toward, should the discussions not meet with the desired result. Besides the economic and domestic policy dimensions, the aspect of foreign policy was also considered. If Germany should find itself lagging behind other countries, it could catch up if needed. The problem of doing well compared to other countries on the international scene has been posed on many occasions, and was a criterion for action for the German government in this instance, too.

In 1979 the counter-alliance went on the offensive yet again. This time it was Kali-Chemie who adapted scientific data from the USA and fed it to the appropriate authorities. The exonerating points cited were the discrepancy between model predictions and actual measurements for certain chemical species in the atmosphere; and an ozone *increase* measured over the period from 1958 to 1976 at an altitude of 32 to 46 kilometres. Hoechst AG invited all the major daily newspapers to a press conference on the risks of CFCs, where the predictions of the extent of ozone depletion were reported to be irrelevant and measures for the reduction of CFC emissions were labelled premature. The UBA reacted with a press statement in agreement with the Ministry of the Interior. In the statement it was conceded that in part there still remained considerable scientific uncertainties, but the theory could by no means be considered disproved. With reference to estimates from the WMO and the UNEP, a long-term ozone depletion of 15 per cent was calculated.

Behind this press skirmish was the counter-alliance's attempt to free itself from the voluntary (non-binding) agreement reached in 1977. At the suggestion of the Minister of the Interior, a passage was included in the press statement explaining that the Council of the European Community, on the initiative of Germany, had charged the EC Commission to draft a resolution which was to achieve a 'significant and state-supervised limitation on the use of CFCs in aerosols'. This was the first public indication that the Ministry of the Interior was capable of tightening its policy.

Interest politics versus advocacy alliances

Why was the advocacy coalition in Germany unable to make any headway in the 1970s? Besides the fact that the UBA, as a subordinate and newly established agency, kept a low profile and failed to start a public debate, the main reason was that the coalition did not have at its disposal the same crucial source of power that the American pro-regulatory alliance had: scientific laboratories (see also Genscher 1995: 134). At the same time, the Hoechst AG, which itself also pursued little or no original research to clarify the facts of the matter,[26] was essentially quicker and more skilful in its mobilisation and rhetorical appropriation of foreign laboratory data. Above all, however, and in marked contrast to the USA, there was no public partisanship among scientists in Germany. There were no speakers for diffuse interests; indeed, even public fora comparable to those in the USA, where the question could be discussed, were lacking. There were no parliamentary hearings or committees and only desultory media attention.

The types of regulation introduced in Chapter 1 suggest that in such a case as this, where a benefit to many is opposed by a few powerful stakeholders, diffuse interests must be represented by public interest groups and their speakers if the asymmetry between industrial interest groups and environmental protection is to be overcome. In Germany these speakers hardly existed. At the time this was not a problem merely specific to CFCs, but rather a problem of environmental policy in general, as the former permanent secretary of the Ministry of the Interior himself declared: 'Environmental policy was lacking not only effectively organised social groups as allies of the administration, but also a sufficient parliamentary power base' (Hartkopf and Bohne 1983: 156). Environmental policy thus requires

> organised allies outside of the government, in order to maintain a fair position within the government. Economic interests have always been well organised and therefore in a position effectively to represent their interests within the government to their best advantage.
>
> (Hartkopf and Bohne 1983: 149)[27]

International pressure on the German government

At the beginning of March 1977, an international meeting of experts took place in Washington at UNEP's invitation. Although the agenda was limited to questions of atmospheric research, the various points of view were made clear on the side. The USA and Canada were in favour of prohibitive regulations even without further scientific confirmation; the Netherlands were leaning in this direction. Great Britain, France, Belgium and the Soviet Union were against regulations. The American representatives also wanted to deal with regulatory questions at a follow-up conference in April, for which the UNEP meeting was to lay the scientific groundwork that should have led to a certain amount of voluntary commitment to the follow-up conference. The German and British representatives resisted this (*New York Times*, 10 March 1977).

In late 1978 the UBA held an international conference in Munich. Germany stuck to its line that damage to the ozone layer from CFCs was considered very probable, but on account of the considerable uncertainties and gaps in knowledge no need was seen for further measures (a ban, for example).[28] The Federal Republic, however, expected a 'frontal attack' from the USA on this point. The conference once again showed the divergent viewpoints: 'There was a terrible row. The French more than anything were absolutely against taking this seriously. The English too, and many others stood in between. The Americans were very much for it' (GEPO 52).

The German government's position was cautious: on the one hand, it wanted to avoid regulations as far as possible, while on the other it did not want to appear in public as being in the ecological rearguard. It would have been reasonable to expect that Germany would leave this middle path if other countries had taken more far-reaching initiatives. In this regard, an American advance was to be feared. Scientific data were insufficient to motivate leaving this middle course during this period. For example, dramatic new data were communicated by the deputy head of the US EPA to the German Ministry of the Interior in October 1979. These were based on the latest NAS report, which alerted that:

- ozone reduction was estimated at 16.5 per cent;
- a 25 per cent reduction in CFC emissions would still lead to a 13 per cent ozone reduction;
- a 7 per cent per year rate of increase for CFCs would lead to a 56 per cent ozone depletion early in the next century;
- the incidence of skin cancer on the basis of increased UV-B radiation would doubtless increase considerably.

According to the estimate on this last point, a 16.5 per cent ozone depletion rate would lead to an increase of skin cancer of 66 per cent, a 13 per cent depletion rate would cause an increase of 52 per cent, and a 7 per cent increase in CFC emissions would prompt cases of skin cancer to multiply two or three times.

The UBA attempted to take advantage of the propitious moment and proposed the medium-term removal of CFCs from aerosols. The Ministry of the Interior, however, demanded only a 50 per cent reduction because the cabinet wanted to stay within the framework of pending EC regulations. The government now expected a solution to the problem through coordination at the EC level. At the time this meant that the EC would be reinstituting what German industry had already accomplished. Thus German industry gained further breathing space.

As a result, it happened that both the definition of the problem and the proposals for solutions were made by scientists in the USA. There were speakers among them who managed to win relevant actors as allies. In a process that lasted less than three years, a political network emerged around them and asserted itself with its regulatory goals in the USA. In the Federal Republic, there was hardly any public discussion and there were no vociferous speakers for binding measures. The voices that argued for precautionary principles were restrained and isolated, while the 'wait and see' position acted vociferously and self-assuredly. The result was an informal agreement with industry that might have led to comparatively low costs.

The media

In comparison with the USA, there were all in all very few media reports in Germany to draw the authorities' attention at the very beginning. The subject never made the front page. The CFC issue was given no public exposure in the first phase of the controversy. In 1975 eight articles appeared in all, warning the public of the danger of CFCs. Only two scientists (Rowland and Molina) were presented as actors who explicitly favoured restrictions on CFCs. *Die Zeit*, 12 December 1975, quoted Mario Molina: 'The only remaining protection from the radiation is an immediate ban on all sprays. We have to stop blowing a million tons of CFCs a year into the stratosphere.' In the decade from 1974 to 1984 there was only one other report of an explicit demand from a scientist. The *Frankfurter Rundschau* quoted Rowland on 30 March 1977:

If we are convinced of the correctness of our thesis and fail to accept the political consequences, it means that economic considerations prevail over the scientific and health aspects of the problem. I am a scientist, but I am also a citizen and I feel my responsibility to the public. If we hadn't acted like this, our findings would have been dismissed by industry as unimportant, and so we experts have to present our case in public.

The science editor of *Die Zeit*, Thomas von Randow, was the only one who backed these warnings:

> I am therefore thoroughly in favour of stopping the flood of sprays in this country, and of examining at least the feasibility of a precautionary ban, so that industry can move quickly to other aerosols or to the good old pump. Certainly that should not lead us to regard as proven the theory that these sprays are attacking our ozone screen, and the theory that a resulting increase in UV radiation on earth will cause more skin cancer in humans than before and will kill aquatic animals. In this regard, scepticism is unquestionably appropriate.
>
> (*Die Zeit*, 18 July 1975)

Randow points out that it was not 'some environmental fanatics or other' who raised the alarm regarding spray-cans,

> but rather competent scientists in serious research institutes. The probability that their concerns are justified is for that reason alone not insignificant. What is insignificant, however, is whether we have aerosol sprays or not. Drawing the correct conclusion from that should not be difficult for any one, least of all for our Ministry of Health.
>
> (*Die Zeit*, 18 July 1975)

The Ministry, however, did not want to follow this logic. Health Secretary Focke indicated that before legal steps could be introduced, it had to be made clear what dangers, if any, were posed by CFC-based aerosol propellants (*Europa-Chemie* (21)1975: 411).

5   The road to Montreal

The international level

At various places in this study it has been argued that the important definitions of the situation and the significant proposed solutions to the problem were crucially dependent on scientific claims. In the 1970s and early 1980s, these consisted of predictions regarding the long-term development of the ozone layer. The less ozone loss was predicted to be, the less pressure there was to do something. Moreover, the prevailing view was that one actor (or a few actors) could solve the problem alone, so long as other actors did not counter such behaviour. The USA had taken measures that led to a considerable reduction in global CFC emissions. Already in the preparatory phase of the Clean Air Act (*c.* 1977), which provided for a ban on the use of CFCs in aerosol sprays, the USA tried to move other countries to similar measures, but this attempt was unsuccessful. Only Norway, Sweden, Canada and Australia followed – all of them insignificant as producers and consumers of CFCs. A representative of American manufacturers did not conceal his disappointment:

> Who followed? Sweden, a major user of CFCs and a non-producer. Norway, a major user, and a non-producer. Canada, a very small user and a very small producer. And Australia, a non-producer and a very small user. Europe, Russia and Asia Pacific did not follow. The rest of the world looked at the US action and said 'The problem is solved.' Because aerosols were indeed the largest single market and the largest single producer and user took action.
>
> (USIN 21)

UNEP: 'World Plan of Action'

On the initiative of the USA, a meeting organised by the UNEP took place in Washington in March 1977, drawing representatives from 39 countries and the EC Commission (Parson 1993). At this meeting a 'World Plan of Action' for the ozone layer was adopted. The plan included the recommendation that the signatories cooperate in the field of scientific research. In addition, the delegates recommended the creation of a Coordinating Committee on the Ozone Layer (CCOL), consisting of experts from governmental agencies and non-government organisations (NGOs), which would take part in the World Plan of Action. The CCOL, whose work was strongly orientated towards results and goals of scientific research, met once a year from 1977 to 1985. Essentially, it was supposed to prepare political decisions and to define research deficits. The chemical industry took an active part in this group.

The first meetings brought no concrete results. The participants had only bound themselves to collaborate in the area of scientific research; not, however, with regard to regulations. By the end of the 1970s in Europe, only Denmark and the Netherlands came out in favour of banning the use of CFCs as an aerosol propellant, while the Federal Republic of Germany, Great Britain and France were against any ban.

At an international meeting in Oslo in April 1980, with the participation of Canada, Denmark, the Federal Republic of Germany, the Netherlands, Norway, Sweden and the USA, restrictions on all CFC production were advocated on the initiative of the USA.[1] The Americans announced that they would hold their production of CFCs at the 1979 level and, if necessary, would take further steps (Cagin and Dray 1993: 223). The European Community (EC) had agreed in March 1980 to diminish the use of CFC 11 and 12 in spray-cans by at least 30 per cent by 1981 (based on 1976 levels), and to freeze productive capacity for these materials at the level of late 1979, because growth in other applications of CFCs could not be ruled out. The USA considered these measures taken by the EC to be insufficient.

At the suggestion of the Nordic countries, April 1981 saw the founding of a UNEP working group whose task it was to prepare an international convention for the protection of the ozone layer. The first meeting of this group took place in 1982 in Stockholm. The moment was deliberately selected: it was UNEP's tenth anniversary. This symbolic date was meant to underscore the importance of the problem and to increase the readiness of the participating nations for action. The Nordic countries – particularly the host country, Sweden – pressed for control measures, but garnered no support from other countries. Japan and the EC were completely uninterested in such proposals. In the USA, once Reagan had assumed office a change in the policy of the EPA had occurred. In accordance with Republican anti-regulatory doctrine, the new boss of the agency, Ann Gorsuch (later Burford), initiated a change in orientation that led to the Americans taking no active role at this meeting; in the next few years the USA neglected the ozone problem.[2]

Only after further changes had taken place within the EPA (Burford was replaced by Ruckelshaus) did the USA again display a more active policy on the international level. In September 1984, they once more seized the initiative in concert with other countries who were in favour of international measures by founding the so-called 'Toronto Group'. Other members of this group were Canada, the Scandinavian countries, Austria and Switzerland. This group proposed a catalogue of measures with the main option being a major reduction in propellant gases. The EC objected vehemently to this proposal. Having come under pressure, it finally made two proposals. On the one hand, a 30 per cent reduction in propellants (relative to the major period of consumption in the 1970s); on the other, the restriction of production capacities. To this political end the EC could appeal to a scientific study that had appeared in Europe. The Belgian scientists Guy Brasseur and A. de Rudder published a model calculation in 1985, in which no ozone depletion occurred as long as the growth rate for CFCs remained less than 3 per cent per year. The background assumptions for the model were consistent emissions, a phase out of aerosols within four years and an increase in other applications calculated first until 1997 and then for an unlimited period. Not until 2034 would such a growth rate have an effect on the ozone layer. Establishing an upper limit on production capacities while simultaneously lowering aerosol production would actually lead to an ozone increase (Lubinska 1985).

Between the EC and the Toronto group a stalemate situation arose; both groups of countries wanted to internationalise their own approaches to regulation, each by calling upon the other to take measures (Parson 1993: 59). One scientist who worked in Europe at the time commented: 'We were in fact for a production cap, the US was in favour of a worldwide aerosol ban which would have had severe consequences for Europe while CFC production for other uses like air-conditioning could further grow in the US' (USAS 41). Both sides attempted to clothe their own economic interests in environmental policy arguments. However, several observers noted that Europe was not very serious about controlling CFC emissions (Breitmeier 1996: 111; Haigh 1992: 245):

> CFC production went very flat at the time, everybody wanted to use their existing equipment, nobody was ready to build any more, until they found out what would happen. This is the discussion between Europeans and Americans. The Europeans talked about a cap on total production. This was a better approach than what the US wanted: regulating specific industries. But the cap they were talking about was one that would have allowed increased production. So it wasn't a serious cap, it was a future cap. One of the key things was that the EPA recognised in the middle eighties that the amount of production was creeping up. It was only more or less true that CFC production levelled out after 1976 and stayed more or less the same for the next decade. But in fact it went down and then it was rising again.
>
> (USAS 16)

Each side, with good reason, wanted to make its solution binding on the rest of the world:

> To be honest, in my view, the failure to agree to a protocol in Vienna in 1985 was due to both the Americans and the Europeans. We can share the blame. The reason was that at that time the Americans wanted to take action on aerosols and we wanted to take action on production capacity control, which

meant in practice that if one of the approaches would have been adopted, one party would not have to do anything.

(EUAD 37)

The economic dimension was much more readily apparent in the stance of the EC than in the American position. The Europeans were trying to maintain the competitive advantage of European industry; in the USA there was an attempt to sell the newly proposed regulations to industry, using the argument of the 'level playing field'. American manufacturers had, in fact, lost market shares to their European counterparts (Figure 5.1). American industry, however, was not exactly enthusiastic about this argument, since its subsidiaries outside the USA (and in Europe) would be affected by international measures. For American industry the distortion of trade was not as urgent as the problem of renewed unilateral measures or the image problems incurred by a consumer boycott.

The Vienna Convention

In this situation the negotiations in Vienna began: negotiations that were meant to lead both to a Framework Convention and a catalogue of measures (protocol). Since the stalemate between Europe and the USA could not be overcome, the goal of a protocol was postponed. The convention passed in Vienna and signed by 20 nations and the EC Commission provided above all for cooperation in the area of scientific research and documentation of CFC production and emissions. Although this Framework Convention obligated the parties to take 'appropriate measures to protect the ozone layer', these remained unspecified (Benedick 1991: 45). UNEP was charged with leading negotiations at the level of working groups, with the goal of reaching a protocol in 1987. These meetings took place in May

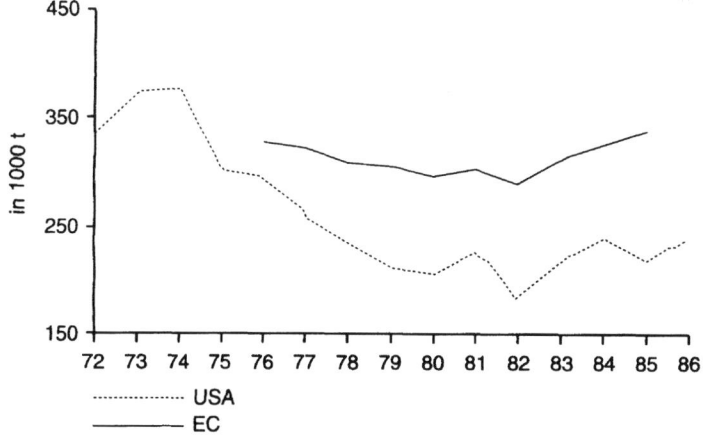

Figure 5.1 Production of CFC 11 and 12, USA and Europe
Source: AFEAS.

and September 1986 in Rome and Leesburg, Virginia. Although the first meeting proceeded largely fruitlessly, the second, in Leesburg, produced several positive results. Environmental groups were admitted to the negotiations, the Soviet Union provided production statistics for the first time, the negotiating partners began to get acquainted and to listen to each other: 'In Leesburg they started to know one another. They started to feel comfortable with each other's point of view and started to listen rather than talk past one another' (Interview with UNEP official).

It was at this time that Du Pont changed its position on the question of regulation, swinging from an unconditional rejection of regulations to a conditional advocacy. As the *Washington Post* reported (10 October 1986), 'The leading producer of chemicals that destroy the earth's protective ozone layer has come out in favor of worldwide production limits'. The trigger for this about-face may have been the confirmation of the Antarctic ozone hole in August 1986 by NASA scientists (Stolarski *et al.* 1986; cf. Chapter 3). Du Pont hinted at this change at the workshop in Leesburg, as one participant remembered:

> In September of 1986, suddenly Du Pont issued their infamous change in position in which they said: We can produce substitutes if there are . . . I think the critical phrase was 'adequate regulatory incentives'. And they went to the Leesburg meeting and the workshops and Paul Halter from Du Pont came to the meetings . . . it was clear that they made a strategic decision in 1986 that some form of regulation was inevitable.
>
> (USEN 31)

For subsequent developments, there were three crucial main factors: first, Du Pont and the business association of the American chemical industry changed their position. Second, the consensus on the old defensive line within the EC began to crumble, because – third – Germany in particular changed its policy. These important changes will now be explored in greater detail.

The USA

The CFC problem had almost disappeared at the beginning of the 1980s; the ozone models from 1979 to 1983 became less alarming, partly because CFC emissions had been reduced. This appearance, however, was deceptive: the reduced use in aerosol applications was more than made up for in other areas, particularly in applications in the fields of plastic foam manufacturing, coolants and cleaning products (Figure 5.2). In 1984 the numbers for world-wide production were once again as high as they had been before the first wave of regulations (Figure 5.3).

The anti-regulatory alliance

The counter-alliance in the USA reconstituted itself at the beginning of the 1980s under the name Alliance for Responsible CFC Policy. President Reagan radically

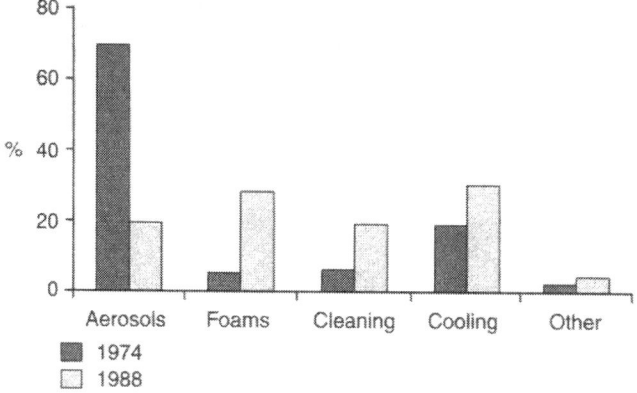

Figure 5.2 Worldwide CFC applications
Source: Du Pont.

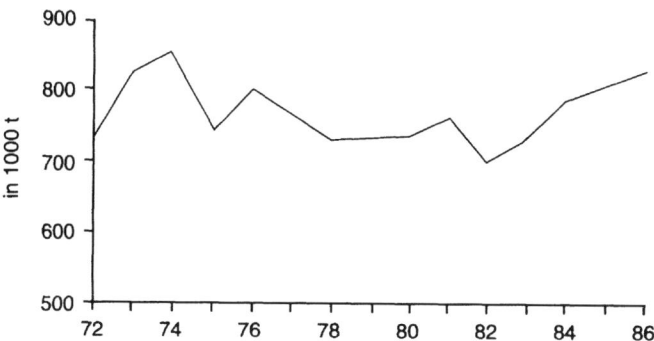

Figure 5.3 Worlwide production of CFC 11 and 12, in 1000 metric tons
Source: AFEAS.

altered the previous environmental policy and made the difficult economic situation the top priority. He saw state expenditures on the environment or fines for businesses as obstacles to an economic upturn. His contempt for ecological demands was notorious, and his opinions on the matter were oft-quoted. He reportedly said 'You and I would live like rabbits if the EPA had its way'. For Reagan, trees and plants were the chief causes of air pollution (Landy *et al.* 1990: 245; McGarity 1991). The EPA budget was cut from 1980 to 1982 by almost 200 million US dollars (from 701 to 515 million US dollars), and over 20 per cent of its staff were laid off. The Council on Environmental Quality (reduced by Reagan from 60 to six members) followed the new economic philosophy.[3]

When the Reagan administration came into power, it had a strong popular mandate to change the role of government in the private affairs of households and firms. Within weeks after assuming the presidency, Ronald Reagan issued Executive Order No. 12291 requiring benefit–cost analysis for all new major regulations.

(V.K. Smith 1984: vii)

According to the Reagan government, the 'decade of the environment' (i.e. the 1970s) had indeed seen many regulations and costs, but no social benefits (Vig and Kraft 1984: 21). The introduction of cost–benefit calculations was meant to facilitate estimating beforehand the effect of all expenditures in the area of the environment.

The new policy towards the CFC problem displaced the clause contained in the Clean Air Act mandating the beginning of a second phase of regulations, should there be further threats to health. The American position was now: 'we have made a major contribution toward solving the CFC problem by reducing world consumption 25 per cent. As a global problem it can now only be solved through international cooperation' (Roan 1989: 87-8). Serious efforts in this area, however, were not initially planned. The new policy served to play for time.

The counter-alliance had its hand strengthened so much in the early 1980s that the advocates of regulation were increasingly marginalised. The early scientific advocates were isolated; even scientists on close terms with them attempted to move them to give up their focus on CFCs. Rowland and an NRDC representative functioned as lonely 'beacons' for the advocates of regulation.[4] The EPA received almost no support for the planned second phase of regulations. Among the 2,300 letters received by the agency, only four registered approval (Roan 1989: 103).

This was largely a result of the reorganisation of the CFC producers. As noted, in 1980 the Alliance for Responsible CFC Policy was founded to engage in lobbying against Congress and the EPA. It orchestrated a great many of the flood of over 2,000 letters to the EPA in order to obstruct the second phase of regulations. In 1981 it opposed the Advance Note of Proposed Rulemaking (ANPR), an announcement of further CFC regulations that dated back to the Carter era.[5] Among others, the Alliance gave the following reasons (Roan 1989):

- hitherto no ozone reduction had been measured;
- the NAS report of 1979 was criticised on many points – the likely prediction of ozone depletion should be halved;
- new studies would lead to a further reduction in the predictions;
- the risk of lost time due to waiting for research findings was nothing to be concerned about.

The 1982 NAS report did indeed cut the predictions by half, to 5 to 9 per cent compared to the previous report:

Although the estimates were close to Rowland and Molina's original 7 to 13 per cent prediction, the tone of the NAS report seemed to indicate that no crisis was imminent. It stated that there was no evidence of a decrease in ozone directly related to human activity.

(Roan 1989: 109)

One year later, the NAS concentrated in its 1983 report on new chemical reactions and their effects on the ozone layer. This report was published in February 1984 and estimated the ozone loss to be only 2 to 4 per cent.

It is significant that the NAS reports have different titles: The 1982 report has 'Causes and Effects of Stratospheric Ozone Reduction', and the 1984 report 'Causes and Effects of Changes in Stratospheric Ozone'. They were backing off a little bit on the word 'reduction'. When you had no long-term reduction then you had no real strong argument for long-term controls.

(USAS 16)

James Lovelock described the state of affairs in the journal *Environment* as follows: 'Had we known in 1975 as much as we know now about atmospheric chemistry, it is doubtful if politicians could have been persuaded to legislate against the emissions of CFCs' (Lovelock 1984: 26). He stressed that increased UV radiation also has beneficial results, as, for example, for illnesses attributable to Vitamin D deficiency or for Multiple Sclerosis. Lovelock asked: 'Now that we are at peace again, it seems worth asking: What were the benefits of the ozone war? Who won and who lost it?' The losers were the small businessmen, the winners the scientists who obtained gigantic sums in funding. Without the 'ozone war' this money would never have been made available.

In the early days of this affair, I was repelled by the unbridled ambition of those who broke every rule of scientific conduct in their mad scramble for fame and funds. The cool excellence of this report suggests that the war was worthwhile, even if it was a messy and gaudy way to gain public support and money for scientific research.

(Lovelock 1984: 26)

One indicator of the changed mood was the decision of the Pennwalt company, one of the major American CFC producers, to expand one of its production plants at a cost of 10 million US dollars. In the period from 1976 to 1982 such investments had practically never occurred (Roan 1989: 110).

The Advocacy Alliance

The Reagan administration, once it had taken office, ignored the Clean Air Act's clause ordering that a second phase of regulations be passed if there should still be some threat. It appeared more likely that the fundamental assessment that CFCs posed a danger would be altered, rather than that appropriate measures would be taken. The NRDC was in contact with Rowland in order to set out the scientific argumentation to back up their petition.

> Rowland had pointed out to me that even with the lower overall estimates of depletion there was still projected to be very significant ozone losses at high latitudes and at high altitudes. So our petition was predicated primarily on that argument from Rowland. It was about redistribution of ozone with unknown consequences. We could not make a direct case for UV-B, but still said: This is a very significant change in atmospheric chemistry. And that itself should be the basis for EPA to take some action.
>
> (USEN 31)

The NRDC also established contact with employees within the EPA, who advised against bringing action too soon.

> Then the Reagan administration took off as if nothing happened. After two or three years NRDC brought a lawsuit to force the action that was supposed to follow from the finding. We were afraid the response of the Reagan administration would be to withdraw the finding that there was danger. But when the first group of appointees left in a cloud of scandal and new, more neutral people were put in, in 1983, we decided to go ahead with the litigation, because we figured they were too honest to revoke the finding that there was danger.
>
> (USEN 31)

The timing of the litigation was discussed within the advocacy network, between the EPA and the NRDC. A former NRDC employee said in this regard:

> But they [the EPA representatives] said to me: Don't bring a lawsuit now. But in 1983 they felt it would actually help them internally if I brought such a petition, so I did. I don't want to give the impression that they were just telling me what to do, but the accurate characterisation of the situation is: We quietly discussed the situation behind the scenes at that time. But they deserve enormous credit since they kept the issue alive.
>
> (USEN 31)

In May 1983 the NRDC sued the EPA for violating the Clean Air Act of 1978. The suit was successful. The agency was ordered to pass new regulations by the end of 1987. The USA was to unilaterally cut back its CFC production and pass trade restrictions against any countries that imposed no restrictions. This seems to have been a means to exert gentle pressure at the international level, although it was not so much an acute threat as a signal that the USA was serious about tightening regulations (Benedick 1991; Brodeur 1986; Parson 1993).

After EPA head Burford was implicated in a political scandal and stepped down in March 1983, the new head, Ruckelshaus, attempted to reorganise the responsibilities for the ozone problem in the EPA. Since the revision of the Clean Air Act the responsibility had lain with the Office of Toxic Substances, which held to the policy it had developed under Burford: no further national measures were provided for unless an international framework convention were to be concluded. On this point there was a controversy within the EPA between the Toxics Office and the Department for International Activities. Ruckelshaus managed to end this quarrel by transferring jurisdiction to the Office for Air and Radiation, a department that regarded the CFC problem as part of the climate problem (its department head was the founder of the EPA's climate programme). The new policy provided for the agency becoming more active again at the international level as well. Before this change the Reagan hard-liners, who also made up the American delegation at international meetings, had surreptitiously taken the position that the United States, if it had to make the decision again, would not repeat the aerosol ban (Roan 1989: 115). The reorientation of the EPA became the official government position – though of course only after some time. From September 1984 until the goals for the negotiations in Montreal were fixed in Spring 1987, the matter was negotiated below the governmental level.

With Ruckelshaus the pro-regulatory alliance received a boost.[6] A mere change of leadership, however, would probably not have been sufficient to pursue the new policy effectively. There were agency staff who had continued to pursue their pro-regulatory goals unnoticed even under Burford, and who were fundamentally involved in the rapid change in EPA policy. They obviously were able to do this because they occupied a relatively low position within the hierarchy. Moreover, they enticed opponents of regulation with an argument that seemed to be industry-friendly: they referred to the competitive disadvantages created as a result of the 1978 regulations, which were strict by international comparison:

> It was easy for them to say internally – even in a Republican Administration – 'Our industry already had to do this, so it is in our competitive interest to make the European industry do this as well'. So there was economic justification to ask Europe to do this.
>
> (USEN 31)

At the turn of the year 1984 to 1985 two people took up posts in which they would exercise crucial influence on further developments. In late 1984 Richard Benedick was named chief negotiator for the USA in the international negotiations in the matter of protecting the ozone layer, and in January 1985 Lee Thomas became head of the EPA. Benedick had been recalled during the lead-up to an international conference on population because he had not represented the American position on abortion vigorously enough (Cagin and Dray 1993: 320f., 395f.). He was transferred into a field that was seen as 'unimportant', and therefore thrown into the ozone controversy, more or less as a punishment. This is not without irony, since it was just this question that in 1987 developed into one of the most important problems in international diplomacy.

Lee Thomas led the EPA further along the course set by Ruckelshaus, working towards drastic international CFC reductions. In November 1986 the EPA policy was cabled to the American embassies. It provided for a two-stage plan: first a freeze on CFC production levels, followed by a long-term phase out. This position was attacked several times within the Reagan administration in subsequent months. The main argument of the opponents of regulation was initially that the EPA and State Department wanted to push this radical policy through in a solo run (Cagin and Dray 1993: 320f.). Benedick, at whom this attack was aimed, had however had this position sanctioned by various other agencies in November 1986; among others, by the Department of Commerce and Energy, the CEQ, NASA, the NOAA, the Office for Management and Budget and the White House Domestic Policy Council. Nonetheless, representatives of the Departments of the Interior, Commerce, Energy and Agriculture, the Office of Science and Technology Policy, and the Office of Management and Budget quickly came to the view that a ruling with such far-reaching consequences actually required a cabinet decision.

The opponents of regulation in various departments began to wake up to the realities in Spring 1987. In late March, a study group of the Domestic Policy Council (DPC) under the chairmanship of the Office for Management and Budget began its consultations. The initial consequence was that the policy for the international negotiations was to be fixed by the DPC until further notice. This meant that at the upcoming round of UNEP negotiations at the end of April 1987 in Geneva no more than a freeze on production levels was to be demanded.

Lee Thomas finally asserted himself in Spring 1987 during discussions between representatives of various agencies and President Reagan, in which the official American negotiating position was determined. Thomas demanded an immediate reduction goal of at least 50 per cent, since he estimated that industry could save about 30 per cent by means of recycling and other measures. In order to provide a stimulus for technological innovations, the figures would clearly have to be higher (Cagin and Dray 1993: 331). In the long term, the chemicals were to be banned. Republican Senator Chafee and Democratic Congressman Max Baucus brought forth similarly worded bills, demanding unilateral regulations in conjunction with trade barriers against products containing CFCs.[7]

Crucial support for the EPA came from Secretary of State Shultz, who publicly backed an international protocol and criticised the Domestic Policy Council for wanting to change the established American position.[8] President Reagan finally approved the policy developed by the EPA and the State Department: the USA would support a reduction of 95 per cent in the international negotiations, but would agree to a compromise at 50 per cent. This formal decision, which only came on 18 June 1987, corresponded to the policy pursued by Benedick.

One of the main reasons that the Benedick–Thomas policy could carry the day was the ineptitude shown by the opponents of regulation. During the consultations in the White House a catalogue of options was drawn up that was finally to be presented to President Reagan for decision. The option formulated by Secretary of the Interior Hodel provided for personal protection instead of CFC regulations: hats, sunglasses and sun-tan cream. The possible increased risk of cancer was to be shifted onto personal decisions, an option that obviously was not tenable. This proposal, thanks to a deliberate indiscretion on the part of the advocates of regulation, fell into the hands of the press. David Doniger of the NRDC heard about these discussions and passed them on to journalists, who at first took them to be a joke and therefore asked for confirmation from Hodel's office. Confirmation was forthcoming not only from this office, but also from the spokesman for the White House and the president's scientific adviser, William Graham. On 28 May 1987 the press reported that Hodel and Graham were against signing an international agreement on CFCs because there were too many scientific uncertainties. The *Washington Post* came out with the headline: 'Administration Ozone Policy may Favor Sunglasses, Hats', and the *Wall Street Journal* led with: 'Advice on Ozone May Be: Wear Hats and Stand in the Shade'. In the latter newspaper Hodel was quoted in the following words: 'People who don't stand out in the sun – it doesn't affect them.'[9] The opponents of regulation were held up to ridicule in the press because of their embarrassing proposal. This was a continuation of the long-running bad press from Irangate to Contragate.

In the resulting public debate – in which President Reagan, who himself had just been treated for skin cancer, took no part – the advocates clearly gained support. The opponents of regulation had made themselves ridiculous. There was nothing left for them but to accept the EPA's ambitious goal. The conservatives may well have comforted themselves with the thought that it seemed highly unrealistic to push this ambitious goal through internationally.[10]

The course of this controversy demonstrates that the network of pro-regulatory advocates (below the government level) could proceed on their own initiative over a rather long period of time. The government had ignored the ozone problem and attached no priority to it in terms of foreign policy. This is also to be seen in the fact that a head negotiator was appointed who had been entrusted with leading the ozone negotiations which supposedly had a low profile. High-ranking staff members from various departments had simply overlooked the importance of the question. By the time they became aware of the explosive nature of the question, it was too late to turn back.

The opposition crumbles

The first sign that Du Pont was rethinking its policy came in September 1986. Joseph Glas, head of Du Pont's Freon Department, explained the company's change of position in a letter to customers. In this letter Glas considered it to be possible that CFCs could be linked with both the Antarctic ozone hole and global ozone reduction. Moreover, it was possible that CFCs could be a greenhouse gas. Since the world wide consumption of CFCs was increasing, it was time to take precautionary measures:

> Because of the new questions and concerns, the inability of science to define a safe, sustainable emissions growth rate for CFCs, and the fact that resolution of these and other uncertainties is not likely in the near term, we have

concluded that it would now be prudent to take further precautionary measures to limit CFCs worldwide while science works to provide policymakers
with better guidance.

(Letter from Joseph P. Glas to Freon customers, 26 September 1986, cit. in
Cagin and Dray 1993: 303)

Contrary to the policy of other producers, it is hinted that it would be possible
to live with regulations, because substitutes would then become profitable. The
Alliance for Responsible CFC Policy endorsed the letter – a complete about-face
for American producers, given that Du Pont had indeed in the previous year
announced another expansion of its CFC facilities in Japan. Previously it had
been argued that regulations could not be justified as long as knowledge was
uncertain; now, however, the very same scientific uncertainty was given as an
argument in favour of regulations.

What were the reasons for Du Pont's turnabout? This extraordinary step gave
rise to a great deal of speculation. In light of the facts available, the following
reconstruction seems sensible:

1   In contrast to other companies, Du Pont had committed itself in public
    statements as early as 1975 to cease production of CFCs should the suspi-
    cions that these substances were dangerous be confirmed. Of course, this is
    not to say that Du Pont would have kept its promise; it is also not clear what
    type of proof would be seen as sufficient.

Indeed, Du Pont defended its position for over ten years, regardless of how the
situation appeared in terms of 'proof'. The commitment, therefore, had no direct
influence on company policy. It provides a publicly effective rationale, however, in
case it should be considered opportune to remember the promise. From 1974 to
1986 Du Pont was the most active actor of the counter-alliance in the USA,
always looking for gaps in the CFC critics' claims. A change in policy, of course, is
made easier if it can be accomplished without too much loss of face. To this end
the public promise was perfectly suited, and the timing was appropriate: the
change took place following the publication of scientific findings.

2   Even relatively 'soft' measures would limit the growth of the field, which was
    not exactly enticing for industry. The director of the Alliance for Responsible
    CFC Policy reduced this to the phrase: 'A business with no growth potential is
    a lousy business to be in' (cit. in Cagin and Dray 1993: 308). Du Pont may
    have speculated that in the event that regulations became inevitable, alterna-
    tive substances would become mandatory for all producers, so that the ques-
    tion of marketability would no longer be so pressing. The important strategic
    question for Du Pont was: How likely are new regulations? And: What
    options do we have compared to other CFC producers?

3   The likelihood of regulations increased with a corresponding change in the
    position of a major CFC producer, since such a step simultaneously robbed

the CFC defenders of two essential resources: on the one hand the alleged impossibility of producing comparable substitutes profitably in a relatively short time, and on the other the resulting implicit admission that the basis for opposition to the Molina–Rowland hypothesis was now gone and that the Molina–Rowland hypothesis would be openly or tacitly accepted. Since Du Pont had waged the worldwide struggle against the CFC critics the most forcefully and became a public symbol in this struggle, its decision would be more important than that of any other producer. Its phase out had to set off a domino effect. If Du Pont had given up defending CFCs, what justification could others then have?

The importance of being Du Pont

In 1986 Du Pont once again took up the research and development in the area of substitutes that it had abandoned in 1980. The expenditures on the newly set-up programme amounted to 5 million US dollars in 1986, 10 million US dollars in 1987 and 30 million US dollars in 1988. By comparison, from 1976 to 1980 3 to 4 million US dollars per year had been spent for research into substitutes (Reinhardt 1989). If these figures are to be believed, then in 1986 Du Pont had no fully developed alternative substances on the shelf. This is a popular ad hoc 'explanation' in the social scientific ozone discourse, which upon closer examination turns out to be a myth.

What was the real status of Du Pont's technological lead? Here a proper degree of scepticism is appropriate. A glance at the chemical industry trade journals demonstrates that all producers of CFCs were confronted with the following problems:

- in no area of application was there 100 per cent substitutability;
- the substances identified as alternatives had to undergo (in some cases long-term) toxicological and ecological testing before they could be permitted;
- appropriate production procedures for the potential substitutes had to be developed (this does not apply for R 22, which was already in use).

The first point casts new light on the popular oligopoly thesis, since in the competition for substitutes there were also competitors from outside of the chemical industry. In applications that had previously employed CFC there were several possible alternatives, including non-chemical products as well.[11] That is, for the CFC producers the chief initial concern was as much as possible to stay in business. Thus it was necessary to wait for the test findings before massive investments could be made in new plants for the production of substitute chemicals. The industry used the interim to build pilot plants in order to test substitutability, technical controllability and customer acceptance. To this end the producers established various research associations in order to coordinate the research. In 1988 it seemed likely that Du Pont would be the first to have production facilities for halocarbon 134a[12] and that it could produce them on a large scale in 1992.[13]

In 1989, the American firm Allied-Signal announced that it would be able to produce F 141b by the end of 1991 in a new plant (F 141b was meant to replace CFC 11 in foams and solution liquids).[14] The German firm Kali cooperated with a British firm in building a pilot facility for F 134a and another substance, F 123. By this time, Pennwalt Corporation came into the market with an alternative halocarbon mixture for freezing. Ironically, however, in 1990 ICI started the first large-scale plant for the production of F 134a. These facts show that the argument often cited – that in 1986 Du Pont was ahead of its competitors in finding alternative substances – must be regarded as a myth.[15]

The producers pursued various options without confirmation of their assumptions regarding profitability. It is unrealistic, therefore, to imagine that there was one producer (Du Pont) who could win a market-dominating position on the basis of a technological lead, and worked toward this goal by means of secret research into alternative substances. Neither in 1986 nor in 1988 was this realistic, and it cannot be confirmed in retrospect.[16] The *New Scientist* reported in April 1987 that the American companies had no technological lead and that all of the major CFC producers had registered patents for the most promising substitutes:

> EEC sources say that American companies, including Du Pont, are ahead in the search for replacements and would benefit commercially from the US's proposals. Du Pont denies this. So does its chief competitor, ICI. 'No one can have a replacement for CFC 11 and 12 on the market before five, or more likely ten years', says Peter Hollins, ICI's business manager for halomethanes.
>
> (*New Scientist*, 23 April 1987: 22; cf. also Umweltbundesamt 1989)

The same article also revealed that ICI, Du Pont and other companies had patented the alternative substance R 134a and that R 22 was developed as early as 1936.

Du Pont, therefore, did not support international regulations because it already had substitutes at its disposal. Economic self-interest no doubt played a role, if in a different form than the thesis of mythical substitutes would have us believe. It was rather Du Pont's long-term strategy which is important here, as both of the following points demonstrate.

In the early 1980s, CFC producers in the USA suffered due to falling prices for CFC 11 and 12, over-capacity and problems of rationalisation. After the first phase of regulations Du Pont had lost one-third of its CFC business.[17] The Du Pont, Allied and Pennwalt companies each closed one of their production facilities for CFC 11 and 12 after the aerosol ban (Reinhardt 1989: 10–12). At Du Pont there were attempts at lowering costs, rationalisation and backwards integration, to be achieved in a new production plant in Texas. Du Pont quickly wrote off this investment; however, the firm was reasonably successful with other measures for rationalisation, so that in 1987 it could become a low-cost producer of CFCs. At the time of the Montreal Protocol, Du Pont gave up a short-term strategy of maximising profits in order not to drive price-conscious customers completely away from the market. The object was to win these customers for a later transfer to alternative substances. 'If we show them, we have a leadership position in alternatives, then they see that as a contribution to their current business', one Du Pont manager said (cit. in Reinhardt 1989: 12). By means of its willingness to phase out CFCs, Du Pont attempted to cultivate the company reputation for acting particularly responsibly and pursuing first-class science: 'The chemical industry in Wilmington has always had a culture that emphasises scientific credibility. They feel they do world class science', as a former NRDC member confirmed. The announcement of its withdrawal from the CFC business was a signal to the customers and to the wider public. Du Pont's rivals could have been tempted to jump into the breach and profit from this retreat. Other CFC producers, indeed, orientated themselves to short-term planning and attempted to gather as much short-term profit as possible with facilities that would likely be obsolete in a short time.

The most important reason for the change in the long-term company policy, however, was the fact that even if attempts at international regulations fell through, the company would have to consider unilateral American regulations or even legal proceedings from skin cancer patients who could sue the polluters for compensation (Roan 1989: 193). In late 1986 the EPA published a study estimating that due to the destruction of the ozone layer, 40 million cases of skin cancer and 800,000 cancer deaths were to be expected in the USA in the next 88 years (*New York Times*, 5 November 1986). As various dermatologists declared at governmental hearings in Spring 1987, the probability of contracting skin cancer in 1930 had been 1:1,500, was today 1:135 and by the turn of the century would climb to 1:90 (Cagin and Dray 1993: 324f.; *Washington Post*, 31 March 1987). The number of new melanoma cases doubled from 1980 to 1989.

At about this time the RAND Corporation devoted a study to the problem of making decisions under uncertainty. The study observes that due to a lack of comparable cases the probability cannot be calculated; as a result, classical risk calculation breaks down. 'The probability must be estimated by policy makers, relying on the best available scientific evidence.' What, however, does the scientific evidence consist of? The author of the study points out that what is needed is not the precise probability of the extent of ozone destruction, but rather only a threshold value or 'cut-off level' regarding the probability of necessary measures. The RAND study argues that this value lies between 30 and 50 per cent.

> If policy makers believe the chance that significant emission controls will be required in the foreseeable future exceeds 50 per cent, adopting additional regulations now appears to be a good investment. If they perceive the chance to be less than 30 per cent, immediate regulations look like a poor investment.
> ('Ozone depletion: probability is all we have', *Wall Street Journal*, 19 June 1987)

The mathematical calculation provides little more than a rule of thumb. The criterion for action is the *perceived probability* of future regulations; but not, however, the 'objective' probability of the dangers. Such a statement can hardly serve as an independent scientific recommendation for action. Politicians, however, could exploit the prestige of RAND in this situation, since the public impression was that RAND, too, was recommending control measures.

If the RAND statement is read as a recommendation to action for CFC producers and the term 'policy maker' is replaced with 'executive board', then it quickly becomes clear in what situation Du Pont found itself in 1986. It was indeed prudent to anticipate regulations, particularly since the probability of measures being taken in the future was at least 50 per cent.

The road to cooperation

In March and April 1987 the representatives of industry no longer had a place in the national delegations of the EC and Japan; and wherever they were still to be found in delegations, they had at least lost control over them.[18] The advocacy network, operating out of the USA, developed an aggressive campaign at the international level via the network of American ambassadors, and sent leading scientists into other countries (among others, to Great Britain and the Soviet Union) to convince them that there was a scientific basis for regulations. American environmental groups, in particular the NRDC, initiated activities in Europe and Japan corresponding to those of the local environmental groups, which had to this point remained largely passive. In Great Britain this was seen as interference in British internal affairs.

> Not until early 1987 did the efforts of some US environmentalists in the United Kingdom begin to pay off in the form of television interviews, press articles, and parliamentary questions about the government's negative policy. Indeed, these American private citizens were so successful that Her Majesty's Government in April 1987 asked the US Department of State to restrain their activities.
>
> (Benedick 1991: 39)

Within the USA the advocacy network (above all representatives of the EPA and the State Department) pushed forward the creation of an ambitious international regulatory initiative. Remarkably, in Spring 1987 the chair of the negotiations, Lang, considered it impossible to achieve more than a 10 to 20 per cent CFC reduction for the next decades (*New York Times*, 28 February 1987; *New Scientist*, 5 March 1987). This should give pause to anyone who thinks, in retrospect, that the CFC case was easy to solve.

By the time negotiations opened in Montreal there were only two remaining negotiating positions: the so-called Toronto Group (among others, the USA, Canada, the Scandinavian countries) and the EC, together with Japan and the Soviet Union.[19] The Toronto Group was in favour of a drastic reduction in CFC emissions by 95 per cent, while the EC wanted a freeze on production facilities (which were, however, not working to full capacity); the USSR and Japan, at first, wanted no regulations of any kind. The developing countries initially showed little interest in ozone regulations but their CFC use was predicted to be increasing sharply in the future.

This constellation left no zone open for negotiation. In the preliminary negotiations to the Montreal Protocol, a stand-off lasted for months between the two largest blocs of CFC producers, the USA and the EC. Since it was the EC above all who would have lost out as the result of an international agreement, it tried to block measures for as long as possible. The critical variable in this process was the change in the position of the EC, which finally agreed to a 50 per cent reduction. Japan and the Soviet Union held back from making any official statements, but gave the impression that they would resist regulations even more firmly than the EC. Little by little the defensive attitude of the EC began to slip: at first it was willing to agree to stabilising production, then to a 20 per cent reduction, ultimately to a 50 per cent reduction.

From December 1986 to April 1987 the situation was not only blocked between the USA and the EC; there was also a stalemate within the EC between the Federal Republic of Germany on one side and Great Britain on the other. While Germany was receptive to the American proposals, Great Britain wanted to give up its old position as little as possible. After a further fruitless round of negotiations in February 1987 in Vienna, the American chief negotiator Benedick described the EC as 'not in a position to negotiate officially' (*New York Times*, 28 February 1987). The German delegate was quoted as a source on the dissent within the EC:

> The American official did not identify the countries that are at odds on the issue, but the West German delegation made it clear that its position was close to that of the United States, and that Britain stood at the other extreme. British delegates declined to define their country's stance and stressed that the community had not reached a common position, as its rules require in such negotiations.[20] (*New York Times*, 28 February 1987)
>
> There were bitter internal controversies, visible even to the press, between the German delegation and the European Commission. 'The Germans are obviously under pressure from their Greens', read the headlines in the English papers at the time. And the Commission wasn't exactly thrilled that a German delegation leader at the civil servant level – not a minister – was talking about the possibility of unilateral initiatives. That was something spectacular at the time.

(GEAD 9)

Progress in the negotiations did not occur, then, until Great Britain in its turn relinquished the presidency of the EC and in July 1987 also left the triumvirate.

Europe and Germany

The EC

Huber and Liberatore (2000) distinguish three phases in the policy of the European Community: In the first, defensive phase (1977–83), it pursued a policy of status quo; in the second, active phase (1984–87), it took part in the international activities that ultimately led to the first binding steps. In the third phase (after 1988), it pressed for the tightening of the regulations already in place. Since 'active' is somewhat of a euphemism – the EC was not prepared to agree to serious CFC reductions before 1987 – one should perhaps characterise the role of EU in this case as developing from a laggard in the 1980s to a (partial) leader in the 1990s (Oberthür 1999).

Opposing positions

The policy of the EC was determined by the European Parliament, the EC Council, the Economic and Social Committee (Ecosoc) and the EC Commission. In general, the European Parliament favoured progressive (precautionary) measures, while Ecosoc supported industry. The theme touched on two different policy areas within the EC, environmental policy and trade policy, in which different rules applied for determining a majority. In environmental policy unanimity was required, whereas trade policy demanded a qualified majority. Not until the ratification of the Single European Act in 1987 did the EC have an environmental mandate (Beutler *et al.* 1993: 510). Besides these institutional factors, it should be recalled that in the international negotiations the EC followed a policy which sought to push through, in addition to the substantial questions, the acceptance of the EC as an independent unit in the international arena. As a result, complications arose in the course of the international negotiations.

In August 1977 the EC Commission put forward a Council recommendation that took into consideration the fact that there were international efforts towards regulation and that voluntary measures were already in place on the national level. CFCs were a known problem, and there was a feeling that something had to be done. The Commission said that on the basis of available data, it was impossible to assess the risk. It therefore suggested controls on the production capacities of the EC, in an effort to avoid major economic or social repercussions. In concrete terms, four actions were proposed:

1   To coordinate research into technical aspects at the EC level.
2   To develop substitutes for CFC 11 and 12.
3   To avoid leakage of these substances.
4   To forestall expansion of production capacities.

The Ecosoc and the European Parliament commented on this recommendation in September 1977. The Ecosoc greeted the recommendation and pointed out that it had no legal force. The considerable uncertainties would suggest a 'wait and see' attitude, particularly since only a slight level of ozone depletion was to be expected. Nonetheless, precautionary measures should be taken, given that the development of substitutes would necessitate a lengthy interim period. Any possible measures should not entail distortions of normal trading conditions within the EC. The European Parliament, in contrast, favoured legally binding forms (i.e. an EC regulation or a decision of the Council) and advocated more extensive measures in the area of aerosols, in order to catch up with the progressive position of the Americans.

Europe takes a cheerful view

In May 1978 the Council issued a resolution on 'CFCs in the environment' that essentially followed the Commission's proposal, declaring that production capacity was not to be increased and that research should be coordinated. According to the decision, UNEP was to be responsible for leading the international process and the producers were to develop substitutes. A year later the Commission presented a new recommendation that took up the findings of the international conference in Munich in December 1978. The aim of the recommendation was a reduction in aerosol usage of 30 per cent (based on 1976 levels). The European Parliament and Ecosoc backed this recommendation; at the same time, Ecosoc went on record as expressly declaring that more extensive reductions were not feasible. In March 1980 a legally binding decision of the Council was passed allowing no extension of capacities and enforcing a 30 per cent reduction of CFC consumption in aerosols. Within a year, a new decision was to be taken in the matter, based on new economic and scientific data. Both measures were implemented without major problems, since they would have occurred in any case on the basis of market developments (recession).

In June 1980 the Commission declared that there were significant discrepancies between the findings of American and British researchers. Whereas the American researchers considered an ozone depletion rate of 16.5 per cent to be probable, the British expressed doubts about the validity of the hypothesis (NASA 1977; UK DoE 1979).

> In Europe the governments did not like the Americans to tell them what to do, so there was some research going on. These governments were hoping that science would find different results than those in the US. So there was always a little bit of tension.
>
> (USAS 41)

At the time the European Commission took the view that CFCs had until now had no effect on the ozone layer. In the following year the findings of one of the scientific workshops organised by DG XII were accepted, findings that emphasised the seriousness of the problem.[21] Both future ozone depletion and new areas of CFC usage were cause for concern. From this, the Commission did not conclude that a change in the current policy was appropriate. The Commission demanded production data from CFC manufacturers in order to maintain control over capacities. The Ecosoc, as before, considered there to be significant scientific uncertainties and therefore felt that further research should be carried out. The European Parliament saw the proposed measures (cap and 30 per cent reduction in aerosols) as a minimum, and wanted in addition to inform consumers in order to reach the goals set by the EC.

After the USA had changed its position in 1983 and supported the proposal of the Nordic countries (a worldwide aerosol ban), the EC lost a powerful ally and had to react. The initial reaction was a new version of the old strategy towards restricting production capacities. As mentioned, this was a planned future restriction on production, since facilities at the time were not operating at full capacity. Nonetheless, a proposed measure usually means to admit that a problem exists. Once the existence of the problem has been conceded, however, it is usually too late to turn back in principle.

Special status of Europe

In October 1984 the Commission embarked upon the international negotiations leading to the Vienna Convention. The first concern was the status of the EC, which wanted to sign as an independent actor, something that other countries greeted with scepticism. The EC pursued recognition as an economic unit (Regional Economic Integration Organisation, REIO) as a political goal.[22] This was opposed by the USA who feared that the sought-after regulations left too much leeway open to member states, since a reduction in total EC production could be compatible with an expansion of production by individual CFC producers, which would lead to a distortion of international trade. The EC nonetheless carried its position through. For the first time the EC was able to be party to a 'mixed' international agreement that did not have to be signed by any of its member states (Jachtenfuchs 1990: 264ff.; Temple Lang 1986). The EC was able to manage this above all because it became clear to the USA that the EC's position was the expression of a common position in Europe, based on a positive coordination (Scharpf 1993). The USA had only two options: either to agree to the EC's position as a whole and to trust it, or to reject it as unreliable. The Vienna Convention was finally signed both by the Commission and several member countries.

In December 1986 in Geneva, new negotiations began towards a catalogue of binding measures, or protocol. The EC's mandate in these negotiations was initially extremely restrictive, since its position had been established in 1980 (a limit on capacities and 30 per cent reduction in aerosols). A wider mandate could only be obtained through a decision of the Council. In the Council, the countries with the largest CFC producers set the tone (particularly France and Great Britain, see Table 5.1). Great Britain held the EC presidency at the time and frustrated any attempts at altering the mandate in the Council. The negotiations could only make progress after January 1987, when Belgium replaced Great

Table 5.1 The most important European CFC producers and their production capacities (absolute and relative to the EC) circa 1980

Country	Producer	Production capacity (t/year)	Proportion in %
Great Britain	ICI, ISC	150,000	21.5
FR Germany	Hoechst, Kali	128,000	18.3
France	Atochem	125,000	17.9
Italy	Montefluos	80,000	11.5
Netherlands	DuPont, Azko	77,000	10.9
Spain	Atochem, Kali	66,000	8.9

Source: Huber and Liberatore 2000

Britain in the EC presidency. As late as March 1987, a member of the American Senate reported that Great Britain and France were not prepared to concede that there was an ozone problem (quoted in Dickman 1987). Afterwards, Great Britain nonetheless remained in the triumvirate (of past, present and future Council presidents) and played an important role at closed meetings among the most important countries; but there, too, they had to leave after a further turnover of office in July 1987. The British delegate was shut out of the consultations for this formal reason in September 1987. 'She was nonplussed when she showed up at the first of these crucial conclaves in September 1987 and was excluded by EC colleagues on these technical grounds', as Benedick remarks (1991: 36). Once Belgium took over the presidency, it was possible to reach an informal agreement in the Environmental Council. The Council gave the Commission more leeway for action, allowing the proposal of a 20 per cent reduction in the production and use of CFCs to be made to the international negotiating partners.

The EC environmental ministers decided in March 1987 on a three-step policy: first, production of CFCs should be frozen at the level of 1987, to be followed by a reduction of 20 per cent relative to 1986. Afterward, possible further measures should be taken, but these measures were not defined more precisely. As Lang (1988: 108) summarised the development: 'Between the experts' meeting in February and April 1987, the EC position began to shift (first automatic reduction at 20 per cent).'

Germany

The advocacy alliance

The first visible steps toward forming an advocacy network in German were made by scientists from the Energy Working Group (Arbeitskreis Energie, AKE) of the German Physics Society (Deutsche Physikalische Gesellschaft, DPG). In January 1986 this working group drew attention to the 'impending climatic catastrophe' at a press conference. The causes for this problem were named as CO_2, nitrous oxide and 'various hydrocarbons' (i.e. CFCs).

As methods of avoiding or limiting the impending climatic catastrophe, the AKE proposes: intensifying climate research; the constant and continuous lowering of rates of emission [of CO_2] by 2 per cent per year; the lowering of energy requirements.

In the following year a proposal by German atmospheric scientist Grassl was published, suggesting that trace gases be divided into three categories and that different measures be taken accordingly. Infrared-active trace gases (that is, CFCs) should be replaced with substitutes, CO_2 should be reduced as much as possible and methane and nitrous oxide emissions should be tolerated.

These warnings caused a furore in the mass media and met with a direct response from politicians. It is noteworthy that the focus was not the ozone layer ('ozone hole'), but rather catastrophic changes in the world's climate. This 'Ur-definition' had a staying power that extends to the present day.[23] The AKE appeal was followed in June 1987 by a 'Warning of Impending Worldwide Climate Changes due to Humans', which was drawn up jointly by the DPG and the German Meteorological Society (Deutsche Meteorologische Gesellschaft, DMG). This warning turned out to be less dramatic; the word 'catastrophe' no longer appeared in the text.[24] In this 'warning', the physicists and meteorologists demanded the replacement of CFCs with substitutes. The scientists called upon Chancellor Kohl, in an open letter signed together with the environmental groups BUND and Greenpeace, to ban CFCs as quickly as possible. In July 1987, Greenpeace started a campaign against the use of CFCs in spray-cans. Here, too, the effects on the climate were given first priority, and only in second place was the effect on the ozone layer mentioned (*Der Spiegel*, 34/1987).

In July 1986 representatives of the ecological movement BUND called upon the German government to immediately forbid the use of CFCs in spray-cans and limit their use in the production of plastic foam. In May Green MPs started an oral inquiry on the topic in the Bundestag and in November they proceeded to a written question. In December 1986 several SPD members tabled a motion in the Bundestag to ban CFCs. Reference was made both to the effect of CFCs on world climate and the ozone layer and to the probability that the ozone hole is caused by CFCs. The demand was 'that the Federal Republic of Germany, by means of national measures, play a leading role internationally as well'. Atmospheric scientist Crutzen stated in *Die Zeit* that he was firmly convinced that the ozone hole had 'essentially anthropogenic causes'. Crutzen, jointly with Frank Arnold of the Max Planck Institute for Nuclear Physics, had developed a theory that, according to the press office of the Max Planck Society, 'conclusively accounts for the origin of the ozone hole' (*Die Zeit*, 13 March 1987).

In the government statement of 18 March 1987, Chancellor Kohl mentioned increasing global threats to the earth's atmosphere and the necessity of national and international measures (Enquetekommission 1990: 209). On 15 May 1987 the CDU (Christian Democratic Union) dominated second chamber (Bundesrat) called upon the federal government to ban the production and circulation of CFCs. A week later, the Bundestag passed a recommendation of the Petition

Committee in which bans and reductions were likewise demanded. The chemical industry made a declaration of self-restriction, according to which the use of CFCs in spray-cans was to be reduced. On 14 October 1987 the Committee for Environmental and Natural Protection and Reactor Security of the German Bundestag conjointly resolved to appoint a Study Commission into 'Precautions for the Protection of the Earth's Atmosphere'. This was preceded by parliamentary initiatives of the SPD and the Greens (Enquetekommission 1990: 558f.).

It should be noted that the CFC problem in Germany entered public discussion via the detour of climate change. The representation of diffuse interests was carried out by actors from science, the media and politics (both officially from Bonn and unofficially from the environmental groups). It was not before these speakers represented diffuse interests that they appeared on the political agenda. Since the official policy dealt with the problem at the highest level, it robbed the 'natural' candidates for the role of speakers for diffuse interests (Greenpeace, BUND, the Greens, and so on) of a basis for successful action. By means of this preventive strategy, the activities of the environmental movement remained largely limited to the role of nipping at the government's heels: it could only demand more comprehensive and more rapid measures, not challenge the basic orientation. In the Bundestag debate on the outstanding ratification of the agreements of Vienna and Montreal, all established parties were agreed that the catalogue of measures from Montreal had to be enlarged. Schmidbauer (CDU) demanded a 95 per cent reduction by the year 2000, Müller (SPD) a total ban on CFCs by 1995; Segall (FDP) a reduction of 90 to 95 per cent within a shorter time, and Knabe (the Greens) demanded a reduction of 95 per cent by 1999 at the international level, and a ban on CFCs within Germany by 1994 (Deutscher Bundestag, 22 September 1988).

Environmental organisations in Germany intervened late, partly because they thought it not easy to reach people with such a difficult-to-pronounce topic as chlorofluorcarbons, partly because they feared that putting the atmosphere on the agenda might give a boost to the nuclear industry (since at the mid-end of the 1980s, CFCs and CO_2 were perceived as coming in a 'double pack'). In Germany at least, the environmental groups had fought their longest and fiercest battles against nuclear power. They were thus anxious to provide any argument for their long-standing enemy. Greenpeace started a campaign against the use of CFCs in spray-cans in July 1987, sending out an SOS to 1.8 million Germans: 'Help to stop the pending catastrophe!' Of course, it also targeted all other aspects of CFC use.

After the Montreal Protocol, Greenpeace was pushing for stricter legislation, calling to close 'loopholes' and to reframe the issue of substitutes which had been largely defined by the chemical industry. In Germany, it proposed and promoted vehemently an alternative technology for home refrigeration ('Greenfreeze') which was not only CFC and HCFC-free, but also HFC free (running on 'Foron', a butane and propane mix). With this, it tried to prevent R 134a from replacing CFCs. Although R134a does not contain chlorine and therefore does not attack ozone, it has a Global Warming Potential. Greenpeace wanted to ensure that we did not replace an ozone killer with a substance posing a hazard to global climate. The 'Greenfreeze' campaign was successful. German industry initially scorned the prototype fridges Greenpeace had produced with the former East German firm Scharfenstein but soon came to take over the technology. Today, all big German producers of home refrigeration appliances use this technology instead of R 134a which is widely used elsewhere in the world. China for its part has adopted the 'Greenfreeze' technology in up to 50 per cent of fridges (personal communication with Benny Härlin and Wolfgang Lohbeck from Greenpeace).

After a mobilisation had taken place in which the American advocacy network took the initiative, factors of domestic policy saw to the further self-reinforcement of the German advocacy network. The SPD opposition pushed for a ban on CFCs. The Greens got a boost in the parliamentary elections in January 1987, increasing their votes from 4.1 per cent to 7 per cent. In April 1986 the reactor accident at Chernobyl occurred, and numerous chemical accidents along the river Rhine had sensitised the public to environmental questions.

In June 1986, the government reacted with the creation of a special Ministry of the Environment which took over several functions from the Ministry of the Interior. This was not an easy task, as Weale *et al.* (1996) have commented,

> given the symbolic need to demonstrate a commitment of a strong policy of environmental protection in the wake of the perceived crisis of Chernobyl, it is not surprising that it was possible in 1986 to overcome the long and well-established bureaucratic resistance, particularly from the Interior Ministry, to the loss of environmental functions.

Returning to the issue of international negotiations, two points regarding the German position have to be made. First, the change in the German position was made easier by the fact that environmental policy reforms had already been prepared under the previous social–liberal administration – they only needed to be implemented (Weidner 1989: 16). Second, the German government was forced to take international environmental negotiations seriously and even to play a leading role; as observers have noted. Lang pointed out that '[s]ince the early eighties the FRG-stance on air pollution, for instance, has changed from reluctance towards international action to a very positive attitude' (Lang 1994: 176). This was elaborated by a member of the German environmental administration in the following way:

> In about 1985 there was a great interest in the government in hooking up with the big international agreements that were being prepared and seeing them as government business too, and not like before, where there were more people in the delegations from industry than there were from the government. They wanted to get that back into their own hands, also in order to keep the government's particular position to speak at international conferences from becoming no more than empty boasting. They felt bound by what they presented in talks with the Americans and UNEP, and then they actually

implemented it. Once the German bureaucratic machine gets going, then it is serious about things.

(GEAD 12)

The Enquete Kommission

In the process, scientific claims were taken up by the Enquetekommission, the Study Commission of the German Parliament, while at the same time a conflict of political goals was solved. In the following passage I deal with the first point, postponing the second until the next section, 'Germany's leading role'.

The Enquetekommission (EK) for the Protection of the Earth's Atmosphere, which like every EK in Germany (and in contrast to the NAS in the USA) contained politicians and representatives of associations in addition to scientists, quickly came to a unanimous judgement. Since this is not normally the case, it can be assumed that this result had been targeted politically.

> The fact of the unanimous vote, in my view, was a dictate of political wisdom. In dealing with a problem of such importance, the democratic powers are well advised if they pass jointly whatever can be kept to jointly – even at the price that they leave certain things that they consider better and more important out of the final vote.
>
> (GEAD 19)

The EK carried out several public hearings, two alone on the topic of CFCs and stratospheric ozone in early 1988. Invited experts from the field of atmospheric science were Brasseur, Brühl, Crutzen, Ehhalt, Fabian, Isaksen, Labitzke, McElroy, Rowland, Stolarski, Sze, Watson and Wuebbles. The industry experts were Bräutigam (Kali-Chemie), Hoffmann (Hoechst) and McFarland (Du Pont). The composition of the group of experts demonstrates the dominance of pro-regulatory advocates.[25]

To be sure, the Commission of Inquiry was officially appointed only after passage of the Montreal Protocol. Already before that, two of the most active policy entrepreneurs were busy working to convert hangers-back at the EC level:

> Once it was clear what our policy was, we purposefully visited the most important countries, and in fact always in such a manner that Schmidbauer [CDU] travelled to the socialist governments while I went to the conservatives. Schmidbauer visited Mitterand, I was in Great Britain, but also in Belgium. We discussed it and agreed on it, not only at the governmental level, but also on the parliamentary level.
>
> (GEPO 52)

The report of the EK had the invaluable advantage that it spoke to the parliamentarians of other industrial countries much more than the position of a government. Influence from government to government is something different from that exercised by parliamentarians on their own government. The report of the EK was discussed in the parliaments of almost all the democratic countries.

(GEAD 19)

That was a pilot project that had gained international attention and was much sooner accepted by parliamentarians than anything that is simply official government opinion, because then right off all they think about is self-promotion. Nobody could say that his political movement had jumped ship; Labour, republicans, Gaullists, they could all say: my people were in it too.

(GEAD 19)

Germany's leading role

Germany faced a conflict of political goals: it pursued a course of both furthering European integration and advocating regulations of similar stringency as the Americans. In about 1986, Germany began to leave the EC common front, which it had essentially determined together with England and France. When the USA noticed this, it attempted to break the Germans away from the EC policy, to no avail. Why did the German government change course and stop listening to its own industry? One possibility is that the German government had entered a regulatory competition in the field of environmental law. Héritier *et al.* (1996) have given various reasons for why a Europe-wide competition could take place here: dissemination of higher environmental standards could spell competitive advantages within the EU. In this case, however, factors of domestic policy above all may have played a role. After the coalition government had become involved in 'green topics' in domestic policy, it attempted to play a leading role internationally as well:

In the run up to Montreal we actually kept to a sequence all the way to the end of the undertaking, a sequence that went: we want to be at the head of the movement, the EU has to be better than the rest of the world and the industrial nations come afterward. During the process of tightening the regulations at Montreal we always set the strictest goals as a nation, we pushed the strictest goals through in the EU, so we were better than Montreal and Copenhagen and London and all the others came behind.

(GEAD 19)

We framed our declaration in such a way that Germany had to do more than the other countries. It was a kind of process in stages, so first Germany, two years later the EU and another two years later the rest of the world. So that relatively credible guidelines were drawn that didn't demand too much of the others. That began already at the end of '86.

(GEPO 52)

The Federal Republic knew how to take advantage of the situation that had arisen from the conflict between the EC and the USA. Germany knew that on the one hand, the USA had an interest in a change in attitude on the part of the EC; and on the other hand, the EC had an interest in a common EC policy. In particular, it was important to the EC, after the passage of the Single European Act in July 1987, to be taken seriously as a single unit internationally. The German policy seized the opportunity here to unite all these different – and at first glance incompatible – goals with its own actions. The USA was denied the satisfaction of Germany following American CFC policy, although Germany was entirely open to strict regulations. Loyalty to Europe prevented this from happening. The EC, however, was denied the satisfaction of supporting its majority line forever; this was prevented by an orientation towards a strict regulation of CFCs. Above all, industry was denied the satisfaction of continuing to be left alone.[26]

> The Americans thought that the German policy was the soft spot. But that didn't work. Then the Americans agreed that the EC could act as a whole. That happened during the preliminary negotiations in Montreal. . . . The EC was clever enough to take advantage of that, because at the time it wasn't really accepted yet; I'd almost say it was a nonentity.
>
> (GEIN 3)

> The American chief negotiator simply never saw the political side of it. It's only because we were working at the European level that we got so far at all, that the EC agreed first to a reduction of 50 per cent, then 100 per cent. That's a result of his still being caught up in the idea that if you make a deal with the Germans or the Dutch, you've achieved something for Europe.
>
> (EUAD 4)

The price that had to be paid for a united European policy was that the EC had to give up its timid approach to regulation. The German government managed to unite both these aspects and to carry out this policy. It should also be noted that a renewed informal agreement with industry initially ran aground; as a result, in 1991 an ordinance banning CFCs and halon was enacted. The political style during this period is clearly confrontational. Not until a year after the ordinance did industry publicly commit itself to end production and use of CFCs by 1994 (Bundesregierung 1994).

6   The Montreal Protocol and after

In the preparatory phase of the Montreal Protocol the participants found themselves in a deadlock: on the one side were countries willing to take action, on the other were countries against. Both based their positions on principled opinions, the precautionary and wait-and-see approach. How could the deadlock between the two camps be overcome? By 1987, industry representatives had lost their status in many delegations as unofficial spokespersons for their respective governments, while representatives from environmental organisations gained influence (Benedick 1991) – a clear sign that well-organised interests were on the decline and that diffuse interests were gaining more weight. When the formal negotiating meeting began on 1 December 1986, there were four environmental NGOs attending, as well as representatives from industry and business. During the second round in February 1987 there were more NGOs attending and even more during the April meetings. NGOs and the media demanded action.

This change reflects the growth of the policy network, which was in favour of strict regulations. Cooperation emerged after the obstinacy of the opponents of regulation was broken and the pro-regulation network was ready to grant them exceptions. A breakdown of the negotiations was the least preferred outcome for the pro-regulation network, but not for the European, Soviet and Japanese industry and its supporters. They would have profited from a continuation of the status quo. The required unanimity conferred an advantage on them through the negotiation system's 'default condition' (Scharpf 1988). Their negotiation strategy was stubbornness: they tried to hold out as long as possible. The pro-regulation network, on the other hand, wanted to gain as many concessions from them as possible. Not surprisingly, the result was a protracted negotiation period.

The fact that it was over after merely 18 months is due to a dramatic gain in the credibility of the pro-regulation network. Three key events helped to tip the balance. First, the discovery of the ozone hole and its symbolic representation changed the perception of the problem completely. As Rowland put it, 'The big loss of ozone over Antarctica has changed this from being a computer hypothesis plausible for the future to a current reality and cause for concern' (*New York Times*, 7 December 1986: E9). Although it was officially not a topic in Montreal, it did in fact have an influence on the negotiations. Second, Du Pont's role as focal actor deserves special attention. As already mentioned, the company had exposed itself most clearly in defence of CFCs and acted as a worldwide speaker for the anti-regulation network. Once it came to see regulations as inevitable, this had a direct impact on other actors of the anti-regulation network, leading to a bandwagon effect. Third, the European Community came closer to the advocates of strict regulation. This change of heart was due to a turn in the German position (see Chapter 5) which was caused mainly by domestic policy events.

The control measures

The catalogue of measures in the Montreal Protocol set a freeze on production totals for the year 1990, a reduction of 20 per cent for the year 1994 and a further reduction by 1999: a 50 per cent reduction altogether of CFCs relative to 1986. Moreover, the Protocol contains the following important clauses:

- a ban on moving production into non-signatory countries;
- a ban on importing from non-signatory countries;
- the signatories are to represent two-thirds of global consumption in 1986: this ensures that the agreement becomes fully legally binding only when the EC, Japan and the USA have ratified it;
- an ongoing scientific re-evaluation.

The compromise provided for exceptions to the regulations for various groups of countries:

- for the EC, which was treated as a regional economic integration organisation (REIO). Given a global calculation of EC consumption, this permitted individual EC countries to hang back in fulfilling their obligations to the same extent that other EC countries met their obligations to cease production more quickly than the Protocol prescribed, provided that all the member states of the EC are also parties to the Protocol.
- for the Soviet Union, which was allowed to keep to its Five Year Plan and complete two CFC plants that were in the process of being built.
- for the developing countries, which were to go unregulated during a ten-year grace period, so long as their CFC consumption remained below a set threshold (Lang 1988: 107; 1989: 109).

The Protocol was signed by 30 parties and took effect on 1 January 1989. These 30 countries were responsible for 83 per cent of worldwide CFC production. The agreement was designed to contain clauses permitting flexible fulfilment, particularly through the obligation of the parties to carry out a regular scientific, technical, ecological and economical evaluation. This was the institutional condition that, from the end of the 1980s, allowed a dynamic to arise that ultimately included more and more substances in the regulations and considerably shortened the timetable for ceasing production.

Integrative bargaining

All these factors led to the hegemony of the pro-regulation position. Once the 'draggers' were isolated, representatives of several key countries were like-minded. They were agreed that some countries should be granted exceptions. Methods of technical problem-solving and bargaining were instrumental in bringing them on board in due time. As we shall see, these methods played an important role not only in Montreal, but also in the negotiations that followed in the years to come. But even more important was a third mode of conflict resolution that could be labelled integrative problem-solving or *integrative bargaining* (Walton and McKersie 1965; Young 1994). As developed in Chapter 1, in contrast to distributive bargaining where negotiators know the shape of a welfare frontier and will therefore 'turn to calculations regarding strategic behaviours or committal tactics that may help them achieve their distributive goals', with integrative bargaining negotiators 'do not start with a common understanding of the contract curve or the locus of the negotiation set' and therefore have a strong incentive 'to engage in exploratory interactions to identify opportunities for devising mutually beneficial deals.' (Young 1994: 100–1). In other words, negotiators here are attempting a comprehensive solution of the problem. The tools used here are much broader than compromises or technical yardsticks that are applied equitably.

The Montreal Protocol really took off as exceptions and adaptation clauses were granted to the draggers. It entailed favourable clauses for nearly all big competitors of the USA. The EC was acknowledged as a regional economic integration organisation (REIO), which meant that it was treated as a single unit that could sign an international treaty. The United States rejected this proposition for a long time. However, the EC was adamant about this: after the Single European Act in July 1987 it wanted to be recognised at an international level. The Soviet Union was allowed to complete two new CFC production plants already under way in its Five Year Plan. Finally, the developing countries were granted a ten-year grace period in which they could continue to produce and consume (specified amounts) of CFCs (Benedick 1991). All these exceptions entailed competitive advantages for America's economic rivals. The fact that the USA not only accepted such a solution, but actively promoted it, can only be explained by the influence of the transnational policy network in favour of strict controls (and not by a 'realist' or 'instrumentalist' reading which would stress the hegemony role of the USA: as we saw in Chapter 5, it took some effort to get the Republican administration going down this path). Its cognitive orientation and normative commitment led to a dedicated search for a comprehensive solution of the problem. The battle to protect the ozone layer was won by the pro-regulation network first and foremost on the domestic ground in the USA. Here, it was possible to break the resistance of the anti-regulatory forces inside the Reagan administration which, in the Spring of 1987, tried to derail the pro-regulation position developed by EPA and State Department for the international negotiations – at a point when industry had given up its hard defensive line.

Technical problem-solving

In negotiating the Montreal Protocol and its subsequent amendments, technical yardsticks (Ozone Depletion Potential, ODP and Chlorine Loading Potential, CLP) were crucial to arrive at common solutions. They allowed the costs of regulation to be distributed in a fair (i.e. generally accepted) manner. The ODP is a weighting system for different ozone depleting substances in which CFC 11 was assigned the (arbitrary) value 1, as Benedick (1991: 78) summarised it:

> On the basis of this weighting system, the negotiators could craft a protocol provision that allowed substances to be treated for control purposes as a combined 'basket' rather than individually. This formulation gave countries an incentive to impose greater reductions on substances that were relatively more harmful to the ozone layer, as well as those whose uses were less essential to them.

This furnished a technical gauge for estimating the destructive potential of different substances and each country's contribution to the problem. By this token, several countries realised that they could achieve the required reduction quota by cutting back in an area that was not vital for its economy. Japan, for instance, initially opposed the draft Protocol since CFC 113 was one of the included substances (Benedick 1991: 79). It was heavily used in Japan's computer industry as a solvent. When Japanese representatives realised that Japan could arrive at the required reductions by cutting back on CFC 11 and CFC 12, an important step to agreement was made.

The indicator of the Ozone Depletion Potential was an important factor at Montreal in facilitating technical solutions and compromises in the negotiating process. It was developed at the same time as the long-term effects on the ozone layer were estimated with the aid of model calculations. However, in 1988 it was clear that substantial ozone losses had taken place. This changed the guidelines for regulative policy. Now it could no longer be a matter of preventing future damage, but rather of taking measures that would effect a recuperation of the ozone layer as quickly as possible. So, after the Montreal Protocol was passed, NASA scientists Prather and Watson devised another method of determining the extent to which critical substances would have to be reduced (Chlorine Loading Potential, CLP). This method is based on the idea that ozone depletion does not begin until critical thresholds are crossed. The cumulative amounts of ozone destroying substances in the atmosphere are the decisive factor here, not the absolute value of a substance. In comparison to the analytic ODP, the CLP is an historical indicator. The pre-industrial chlorine content of the atmosphere was estimated at 0.6 ppb (parts per billion) with methyl chloride being the only possible natural source. Today, this accounts for only one-fifth of all industrial sources (Graedel and Crutzen 1995). The Antarctic ozone hole began to develop in the late 1970s, when the global average concentrations of chlorine had climbed to 1.5 or 2 ppb (today they are almost

4 ppb). Here the advocacy coalition began to work towards tightening the Montreal Protocol.

> A logical benchmark for evaluating future control strategies was the return of atmospheric chlorine concentrations to no higher than 2 parts per billion – roughly the chlorine loading at which Antarctic springtime ozone levels had begun to drop sharply in the late 1970s.
>
> (Benedick 1991: 130)

On the basis of the Montreal Protocol it was still possible for chlorine concentrations to grow dramatically – to 11 ppb within one hundred years (see Figure 6.1). The report of the scientific working group of UNEP, *Scientific Assessment of Stratospheric Ozone: 1989*, was taken as a fundamental scientific finding in the political decisions, which is why it was drawn up to synchronise with the international decision-making process (WMO 1989: vi). The most important findings were:

- the Montreal Protocol did not take into consideration that ozone depletion had already occurred in the Antarctic;
- even if the measures of the Montreal Protocol are implemented, chlorine concentration will increase well into the next century, leading to massive ozone reductions;
- the Montreal Protocol was based on model calculations that made no use of heterogeneous reactions. These reactions were first observed in the Antarctic processes, but they can also appear in more temperate latitudes. The ozone depletion rates could thus, despite the Montreal Protocol, be greater than originally assumed;

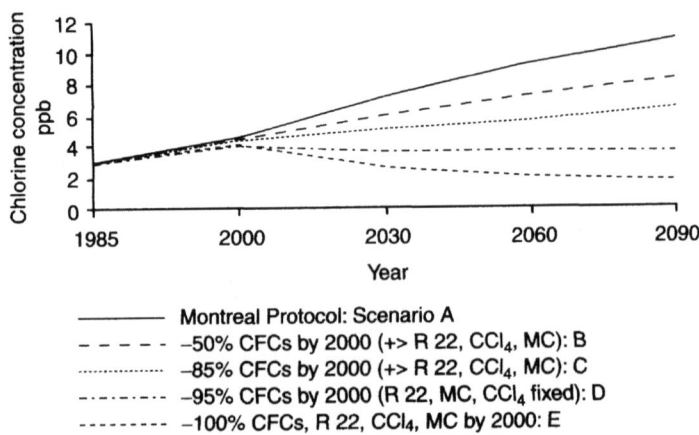

Figure 6.1 WMO/UNEP scenarios

Source: WMO 1989.

- in order to return to natural chlorine concentrations, all fully halogenated CFCs, all halons, carbon tetrachloride (CCl_4) and methyl chloroform (MC) must be abolished and partly halogenated CFCs must be investigated.

As Figure 6.1 shows, only Scenario E leads to a reduction in chlorine concentration to 2 ppb by the middle of this century and thus to a probable disappearance of the ozone hole. In this case, the production of all CFCs must be halted by the turn of the century; however, the production of alternative substances (R 22), carbon tetrachloride and methyl chloroform must also be forbidden to continue. If all critical substances except for CFCs are permitted to increase into the year 2000 and then held constant, then there would still be a massive increase in chlorine concentration, with unknown consequences for the atmosphere.

Bargaining

The readiness of the pro-regulation network to grant exceptions to laggards and the identification of technical measures were instrumental in resolving remaining questions. These were typically contained within brackets of the draft treaty. Main areas of contention concerned the sharing of the costs of regulation. An important issue was the identification of a baseline from which reductions would be calculated. Again, differences between Europe and the USA showed up. One side favoured a formula based on production figures; the other, a formula based on consumption figures. Both positions reflected economic interests. As mentioned before, the United States and other countries initially opposed treating the EC as a single unit. The problem was resolved in a way that gave the EC an advantage. However, the United States still refused to allow EC members to swap production quotas. Under such a rule a production decrease in one country could be compensated by an increase in another. The solution was a compromise. The EC was treated as a unit for purposes of consumption, but not for purposes of production (Cagin and Dray 1993: 335). Level playing-field arguments regarding CFC production became much less important in the years to follow. After all the big CFC producers switched to alternative substances, the London amendments to the Montreal Protocol of 1990 (see following section) extended the unitary treatment of the EC to production purposes as well (Interview, EUPO 4). A similar compromise was found in defining the final reduction goals. The final agreement reached in Montreal was a freeze at 1986 levels, starting in 1989, followed by a 20 per cent reduction by 1994 and a further 30 per cent reduction by 1999. The long-term goal of 50 per cent amounted to a quasi-arithmetic mean between the initial opposing figures of the EC and the USA (20 per cent versus 90 per cent).

Today, there is a vast participation of Southern hemisphere countries in the Montreal Protocol. During the period of establishing the Montreal Protocol, these played a relative minor role. Since both the problem and the scientific research originated in the Northern countries, they took no great interest in the issue. Some countries, like China and India, saw the Montreal Protocol as inequitable and refused to sign it. Although their CFC production at the time was minimal, they were expected to consume 30 per cent of the world production by the year 2000. Both demanded adequate financial support and clearly defined access to alternative technologies before signing up (Wood 1993; also see below, the role of the Multilateral Fund).

The Montreal Protocol: a precautionary treaty?

Parson (1993: 60) points out that from a purely scientific point of view there was no justification for the 50 per cent solution at Montreal. Either a much greater reduction should have been the result, or none at all. The 50 per cent solution, indeed, seems a clear indicator that the result can be traced back to a negotiated compromise in which the obvious solution in the middle (Schelling 1960) was chosen; almost the exact arithmetic mean of the opening positions of the USA and the EC (95 per cent versus 20 per cent reduction), and a compromise in the face of the still uncertain scientific evaluation of the problem.

Even if at the time of the treaty signing in Montreal not all the parties to the agreement were yet convinced of a comprehensive solution, the advocacy network, led by just such a conviction, managed to neutralise the potential hangers-back and ultimately even to win them over. The 95 per cent reduction that had originally been demanded for Montreal was ultimately achieved in the London amendments to the Montreal Protocol.

This success of the advocacy network during the negotiations can be traced back to the combination of an official and an unofficial rationale, both of which mobilised scientific resources. The official position emphasised its precautionary character, while alarm bells were sounded behind the scenes. Officially, a uniform scientific basis for the Montreal measures was created by coordinating different scenarios regarding future global ozone depletion. Leading modellers met in April 1987 at the behest of UNEP at a conference in Würzburg and came to an agreement on the various methods and parameters of their models (UNEP 1987). The Würzburg meeting was the advocacy coalition's attempt to refute the arguments of their opponents. The latter argued that the different models came to different results. Moreover, Soviet scientists claimed that there were no anthropogenic effects on the ozone layer: 'The Soviets ... had to be convinced at that meeting that there is a man-made effect. Before that they were insisting that there was no effect since they didn't want any control' (UNPO 20).[1] Another point was the question whether, besides the ozone hole, there was measurable damage to the global ozone layer. Ten days before the negotiations began in Montreal, Bob Watson attempted to establish a connection with long-term global ozone trends. The *New Scientist* quoted Watson to the effect that a global ozone decline of about 3 per cent had occurred, almost certainly attributable to CFCs (*New Scientist*, 3 September 1987: 24).

Unofficially, the concern regarding the Antarctic ozone hole, whose explanation was at the time still unforeseeable, was skilfully used. This concern led to the view among an increasing number of the negotiating partners that there was no more time to be lost. The unofficial data (from both Antarctic expeditions and from the OTP, which were not published until 1988) provided a much stronger and generally comprehensible signal than the official data (i.e. the Würzburg findings). It was less a guaranteed scientific explanation than a shock that was decisive for the regulations; 'It wasn't a matter of subtle interpretation, this was a sledge-hammer' (USAS 8). The willingness to come to a comprehensive solution was created by the skilful participation of various actors from the advocacy network in the international negotiating process. They managed to maintain a delicate balance between the state of scientific knowledge and the regulatory process. On the one hand, they attempted to separate the political process from the findings of the field experiments (Antarctic expeditions), since the possibility that there was a natural cause of ozone depletion could not be completely excluded. This strategy was forced on them after the first Antarctic Expedition of 1986, which provided no clear evidence of the causal role of CFCs. On the other hand, unofficially they attempted to bring to bear every possible indicator that pointed to the role of CFCs in order to influence the political process. The rationale is obvious: if the dramatic phenomenon over the Antarctic is connected to CFCs, then the pressure to act is considerably greater compared with the model predictions of long-term future global ozone trends, which were much less dramatic. Even if the situation were reversed, the logic still applies. If it turned out that the Antarctic phenomenon was not connected to CFCs, then there would be independent legitimation for the agreed-to regulations. The optimum strategy was, therefore, both to find an official rationale for precautionary measures (in addition to unofficial 'motivation'), and to keep the option open to tighten the measures, should new scientific data justify it. This strategy kept an agreement based on an anthropogenic future ozone reduction in the middle latitudes separate from the possibility of a naturally caused Antarctic ozone hole, but it allowed more stringent measures to be taken in the event that the anthropogenic origin of the ozone hole was confirmed. Thus an important one-way mechanism was created, ensuring that any changes in the catalogue of measures could only be more stringent.[2]

Everyone questioned in my investigation (except for defenders of the official version[3]) admits that the Antarctic ozone hole was absolutely crucial in reaching the compromise. One participant said of the beginning of the conference:

> I know that Watson made presentations in Montreal, showing pictures of the ozone hole. Even if it was decided not to use that information. It is like in a trial when someone says something and the judge will say 'We'll have that stricken from the record'. But the jury heard it. You can strike it from the record but it's in their brain.
>
> (USAS 17)

> I remember . . . B. came when we were in the line to pick up our badges. He said: We have new information about ozone. And it's not good news. So that was the NASA information. . . . So when the Montreal Protocol was signed,

they didn't take into account the ozone hole – at least not explicitly, but they did implicitly.

(USAS 41)

Before Montreal and in Montreal there was no agreement of the big countries on a position. Everyone said: Maybe regulations are too costly, and are we really sure about the risks? But when it was documented that ClO was really existing in enormous big amounts in the vortex, then things started to fit together.

(UNAS 2)[4]

A NASA scientist used the time-lapse photographs made by the TOMS satellite to prepare an animated colour film that clearly showed the growth of the ozone hole in the 1980s. It became popular through many television broadcasts, and was clearly also shown unofficially during the negotiations. Although there was still no scientific evaluation available regarding the cause, several indicators pointed to chemical processes, as, for example, the results of ground measurements from the first Antarctic expedition in 1986 and the ClO measurements taken from the air simultaneously with the negotiations in Montreal.[5] Moreover, no one knew how the ozone hole might develop further: whether it would spread more widely from year to year and reach global proportions or stay a local, perhaps even a transitory, phenomenon.

The shock of the ozone hole undermined and weakened the position of those countries that had opposed regulation, and thus led to their neutralisation; above all because they had committed themselves to scientific reasons for their opposition. Cognitive uncertainty combined with dramatic alarm signals made the negotiating position of the anti-regulatory countries a precarious one. They saw themselves as forced to give way, albeit not completely. They agreed to an arithmetic negotiating compromise that seemed feasible to them, particularly since, ironically, it could be achieved quickly through reductions in applications using aerosols. In order to fulfil the Montreal agreement, the Europeans finally took a measure that had already been proposed to them by the countries of the Toronto Group, which they had previously rejected. The Toronto Group and the progressive EC countries for their part agreed on the 50 per cent compromise and the exceptional regulations, although they actually favoured more stringent measures.[6]

A comprehensive solution

The disposition towards a comprehensive solution to the problem arose from a combination of strong cognitive orientation and weakly marked special interests (cf. Chapter 1), in which the pro-regulation network valued a comprehensive solution (common good orientation) more highly than the pursuit of partial interests. Cognitive orientation of participants can break through roadblocks that cannot be opened in a pure bargaining situation; the case of the Montreal

Protocol is an impressive illustration of this. In the preliminary phase, the negotiating parties found themselves in a deadlock that was led, by way of comprehensive problem-solving, to a technical solution, and ultimately to a bargaining solution. Bargaining led to the identification of a compromise acceptable for all only when the central actors had agreed on a comprehensive solution (extensive CFC reduction). No matter how one looks at it, the crucial question is how the orientation towards comprehensive problem-solving came about.

Diffuse interests became considerably more significant than industrial interests. As already mentioned, around 1987 the negotiating delegations of many important countries replaced the representatives of industry in their ranks with representatives of environmental interests. For representatives of diffuse interests it is rational to gain as much advantage as possible from concessions made by their opponents; breaking off negotiations would not have been rational for them, though it would have been for the other camp, who had profited from the status quo. It is therefore not surprising that the advocates of tight regulations granted their opponents exceptions to the regulations and flexible deadlines. These exceptions applied to almost all of the major competitors of the USA: the EC was recognised as a regional economic unit; the Soviet Union was permitted to finish building two production plants for CFCs, since they had been provided for in their Five Year Plan; and the developing countries were given a ten-year grace period in which they were able to produce and consume CFCs within certain limits. All these exceptions gave the economic rivals of the USA competitive advantages. Thus it is clear that Montreal was not a pure *bargaining* situation, as otherwise the USA could not have agreed to concessions to the 'rest of the world'.[7] The fact that they did so can only be explained by the cognitive orientation towards seeking a comprehensive solution. It can be assumed that the advocacy network knew it possessed symbolic resources that could be brought to bear after Montreal. This would explain why it was ready to make concessions in Montreal. For one thing must not be forgotten: in the balance, the opposing parties profited from the exceptions that were granted them in Montreal. The advocacy network's strategy in Montreal can thus be described by the formula: one step back, in the knowledge that soon it would be possible to make two steps forward.

To sum up: the real stumbling-blocks were removed once the pro-regulation camp made large concessions to countries unable or unwilling to bind themselves to stringent controls. However, this could only happen in a reflexive process that went back and forth between normal bargaining, integrative bargaining and technical problem-solving. Negotiators had to explore which regulations might be feasible and how the burden of the costs would be distributed. Such compromises were contained within 'brackets' of draft treaties. Informal consultations were essential to reduce the number of these brackets and the number of draggers. As agreements between key actors emerged, more and more contentious issues were resolved and more and more draggers isolated. Once such a dynamic set in, even the most adamant parties found it difficult to resist a compromise.

Montreal and after

Interestingly enough, after the adoption of the Montreal Protocol the USA lost the international initiative. To be sure, advocates of regulation pressed for more comprehensive measures – a year after the Montreal Protocol, EPA chief Lee Thomas considered the 50 per cent reduction to be insufficient. He demanded the total elimination of CFCs and, moreover, the regulation of methyl chloroform and carbon tetrachloride. He was supported by David Doniger (NRDC) calling for the USA to take unilateral action. The official advance in this direction, however, came from the EC and UNEP. A few weeks after the agreement to the Montreal Protocol UNEP director Tolba seized the initiative by setting the first follow-up meeting of the parties to the Protocol for 1989, since in his opinion new findings demanded tighter measures. The analysis of the second Antarctic expedition of 1987, which was officially released a few weeks after Montreal, confirmed the hypothesis that the Antarctic ozone hole can be attributed to the effect of CFCs. Once the groups of experts had taken up their work, important changes in position took place in several countries. In 1988 Germany was supported by the British government, as Prime Minister Thatcher began to advocate strict measures.

In 1988 Du Pont announced its intention to cease production of CFCs (without committing itself to any time frame). This move was explained by appealing to the corporate identity of the firm as a concern motivated exclusively by first-class science. The decision was painted as 'a result of pure, hard, cold science making its points in a company where . . . science has always mattered as much as business' (*New York Times*, 26 March 1988). This statement also had the advantage of permitting them to say that they had acted as soon as the science became clear – a defence that could be useful in the event of future legal action on behalf of skin cancer patients. Du Pont had reconsidered its position only after the publication of the OTP's findings on 15 March 1988 (see Chapter 3). A mere three weeks before, in a written reply to three senators who were pressing for regulation, Du Pont chairman Heckert had maintained that there was no available scientific evidence to justify such dramatic measures (Roan 1989: 229).

The view of the Montreal Protocol as a mere first step that had to be augmented by further steps as soon as possible was disseminated everywhere throughout the transnational pro-regulatory network by 1988: among the actors in UNEP, in NASA, in the EPA, in the Enquetekommission and, after Great Britain's change of tack, in the EC as well. The change in the British position was described by the phrase 'The Greening of Margaret Thatcher'. This meant a strengthening of the pro-regulatory network on several fronts: first, it won a new ally with the British government's change of sides, while second, it took an ally from the alliance against regulation. Moreover, it managed to eliminate one of the bitterest opponents of regulation (ICI) at the European level.

In England Margaret Thatcher watched the Green Party in Germany and thought: This can happen here. So she tried to steal the leadership. By the time of the revision of the Montreal Protocol, she convened the first meeting after the Montreal Protocol to strengthen it.

(USAS 5)

In March 1989 the EC Council endorsed a complete phase out of CFC production, favoured by the Federal Republic of Germany and by Great Britain ('as soon as possible, but no later than 2000'). At this point American President Bush gave the new EPA chief, Reilly, the authority to endorse this line. The amendments to the London Protocol were extended correspondingly.[8] At an international ozone conference in London convened by Thatcher, representatives from more than a hundred countries expressed the same idea; developing countries emphasised their need for financial and technical aid in order to move to alternative products. The first meeting of the parties to the Protocol took place in May 1989 in Helsinki, where approximately 80 countries signed a declaration demanding the phase out of CFC production by the year 2000. The interim reports of the various expert groups played their part in this. The scientific group focused on scenarios that sketched the connection between chlorine concentrations and ozone depletion. The technical panel came to the conclusion that at least a 95 per cent reduction in CFCs would be possible by the year 2000. The chances of reducing other materials (like methyl chloroform or carbon tetrachloride) were judged similarly good. The technical and economic working groups demonstrated that short-term substitutes existed for most applications.

The UNEP began negotiations for a revision of the Protocol on this basis. In the course of seven sessions altogether between Autumn 1989 and Spring 1990, it was unanimously agreed to strive to phase out CFCs and halons. With the exception of France, no representatives of industry had been sent in other delegations. The main problem was now the financial and technological support programme for the developing countries. As in Montreal, so too in London unity was reached only at the last minute. This was due above all to the problem of the developing countries. A fund was created whose absolute size remained undetermined, but which was to amount to about 200 million US dollars for a transitional period of three years. The main opponent of such a fund was initially the USA, which only gave its approval after being given assurances that this would not set a precedent for other problems, such as climate. At the London conference, India and China announced their readiness to ratify the Montreal Protocol, including amendments, if the fund was created. In the USA this question provoked a renewed battle between advocates and opponents of regulation. The EPA supported the Europeans and the developing countries in their push to create such a fund. Resistance came from the White House and the Office for Management and Budget.[9] Industry, however, as well as Republican (and Democratic) senators supported the initiative to establish an aid fund.[10] President Bush isolated himself both domestically and internationally with his hard-line policy.[11] The press mocked: 'Penny wise on ozone?' (*New York Times*, 16 May 1990). Ten days before the negotiations began in London, the Bush administration at last announced its agreement. In this case, too, the pro-regulatory network had created conditions below the government level that the American government, if reluctantly, finally accepted.

The implementation of the international agreements took place without difficulty in most countries; in fact, the agreements were even surpassed. Outstanding in this respect is Germany, which considerably shortened the statutory time limits for phasing out all the substances in question. In the USA, Du Pont announced that it would cease production of CFCs in 1994 and halons in 1996. With regard to partially halogenated CFC substitutes (HCFCs), which have only a slight potential for ozone destruction, but in return have a great potential for creating the greenhouse effect, the main producers have taken different paths. The USA advocates a phase out between 2020 and 2050, while in Germany a ban is already planned for 2000.[12]

The London amendments, 1990

British Prime Minister Thatcher convened an international ozone conference in London in March 1989 at which more than 120 governments participated, more than double than in Montreal. To many this appeared to pre-empt the first meeting of the parties in Helsinki (scheduled one month after the London conference), but it turned out to be helpful. More than 90 environmental organisations were present, as well as the international media, all demanding stern action. The following meeting in Helsinki produced a non-binding document calling for a phase out of CFCs 'as soon as possible but not later than the year 2000' (Soroos 1997: 165).

In June 1990, the second conference of the parties was attended by 55 parties and 44 non-parties in London. Thirty-four industrial groups and 14 environmental groups were present. Shortly after the signature of the Montreal Protocol, it had become clear that there were huge ozone losses due to ODS which had not been predicted by any scientific model. This changed the perceived task of regulatory action. It was no longer a question of preventing future damage, but to establish controls in order to achieve the recovery of the ozone layer as soon as possible. NASA scientists introduced the indicator 'Chlorine Loading Potential' (see page 167).

The WMO/UNEP scenarios based on this indicator suggested phasing out all CFCs at the turn of the century, with no substitutes (HCFC 22), no carbon tetrachloride and no methylchloroform being manufactured. This formed the basis for the London agreement where delegates also agreed to establish a special fund (Multilateral Fund, MLF), albeit on an interim basis, to help developing countries to comply with the Montreal Protocol; 240 million US dollars were allocated for the period 1990–93.

The Copenhagen amendments, 1992

Two years after the London amendments, the fourth meeting of the parties to the Protocol in Copenhagen dealt above all with three questions: the Multilateral Fund, the regulation of additional substances and the compliance mechanism (see Gehring 1994: 302–20). The establishment of the Multilateral Fund in London led to an increase in the number of parties to the Protocol, including critical countries such as India and China. Thirty-nine country-specific aid programmes for developing countries were elaborated. As mentioned, the compromise reached in London had planned 240 million US dollars for the interim fund. The donor countries, however, above all Russia and France, did not fulfil their obligations to pay. Russia was suffering from insolvency, while France had reservations about the construction of the fund. The parties recognised the internal economic difficulties of the countries of the former Soviet Union, but at the same time they saw the danger of unilateral non-compliance. On the one hand, this could lead to a downward spiral; on the other, non-compliance would cause deficits that could lead to a non-observance of the aid programmes and finally to an (involuntary) breach of the commitments to the developing countries. This problem was solved by releasing several Eastern European countries from their financial commitments *in hard currency*; instead, they were to provide equivalent service in the form of technology and support. It was crucial that this problem was solved within the framework of the international regime in order to avert the danger of a unilateral withdrawal.

Several Western European countries, among them France, were discontented with the Fund, which was supposed to operate independently of other institutions of the north–south transfer. They wanted to transfer the Multilateral Fund into the Global Environmental Facility (GEF), which had been established in 1992 by the World Bank, UNEP and UNDP as chief financing mechanism for the climate convention – on the initiative of France (Thacher 1992: 199). In London, the USA had likewise objected to the establishment of an independent fund, but now backed it and had a seat on the Executive Committee. In view of this conflict of interests, the EC attempted to delay the decision on the final design of the fund, but this attempt foundered on the resistance of the developing countries. The consensus to strive for a comprehensive solution to the problem was stronger than the national interests on the part of the developed industrial countries.

Before the meeting in Copenhagen, leading manufacturers and consumers of ozone destroying substances had announced their intention to stop production of CFCs by the end of 1995. The corresponding adjustment in the text of the agreement presented no problem. The inspection of additional substances was another matter. Austria, Norway, Sweden and Switzerland were in favour of strictly limiting the production and consumption of 40 partially halogenated CFCs (HCFCs) with the support of the EC. In addition, they proposed a total phase out of these materials for the time period between 2005 and 2010. The USA wanted a guarantee that these materials could also be used after 2020, because air-conditioning systems in current use have a service life of up to 40 years. Here, too, a gradual phase out was agreed to, with production to end in the year 2030 (Brack 1996: 15). HFC, which contains no chlorine but has a greenhouse effect, was not entered into the catalogue for regulation under the Montreal Protocol. This will be a task for a Climate Change Protocol.

The fourth meeting of the parties was attended by representatives from 87 countries in November 1992 in Copenhagen. They agreed to advance the date for phasing out the five substances already under control of the Montreal Protocol, to include other substances and to clarify the status of the MLF. For the first time, hydrochlorofluorocarbons (HCFCs) were included on the list of regulated substances. As already mentioned, northern countries committed themselves to phase out their consumption by the year 2030. Some wanted stricter timetables since alternatives existed for all uses (for example, hydrocarbons for refrigeration). With regard to methyl bromide, some developing countries and Israel were opposed to restrictions because it was critical to their economic development. It is used mainly in the fumigation of soils. Interestingly, they referred to scientific uncertainties about the impact on the ozone layer to justify their opposition. The only agreement that could be reached was that developed countries would freeze production at 1991 levels by 1996 (Tolba 1998).

The issue of the MLF was taken up again. Developed countries agreed to increase the level of funding and to make the Fund permanent. Developing countries rejected all attempts to shift the task to the Global Environment Facility (GEF), since this was seen by the South as being dominated by the rich countries. The level of funding was increased to between 350 and 500 million US dollars for the period 1994–96.

All non-developing countries have to contribute to the fund. Initially, some argued that the contributions to the fund should be based on actual CFC consumption. However, this would undermine the flow from donor countries as they were phasing out CFCs. Therefore, the United Nations scale of assessment was chosen, adjusted to account for those Article 5 countries which contribute to the UN but not to the fund (Biermann 1997).

Interestingly, the agreements reached in Copenhagen were influenced by scientific findings which indicated that ozone loss was much larger than had been predicted, especially over the Northern hemisphere. However, it must be said that this information was not well founded at the time (which several atmospheric scientists later admitted). Senator Al Gore coined the phrase 'ozone hole over Kennebunkport', where President Bush and his family spent their holidays. On *Time* magazine's front page one could read: 'Vanishing ozone: the danger moves closer to home.' However, the prediction about dramatic ozone losses in the north were not confirmed. In April, NASA reported that the depletion was only 10 per cent. This is not to say that ozone depletion has not worsened over time. The important point is that scientific information is fed into the policy process at crucial moments in the negotiation process. Clearly, dramatic news enhances the probability of getting tighter controls. However, this tactic may backfire if the message turns out to be exaggerated.

The Vienna amendments, 1995

Representatives from 149 countries met in Vienna in December 1995 for the seventh meeting of the parties to the Montreal Protocol. Two basic issues were dealt with: the broadening of the scope of the Montreal Protocol, again tackling substances like HCFCs and methyl bromide, and the problem of non-compliance. In Vienna, the phase out date was brought forward to 2020 (with the exception of a small amount of production for servicing purposes which may continue until 2030, cf. Krueger and Rowlands 1996; Soroos 1997).

Methyl bromide is another substance that had been regulated in 1992. While an assessment report suggested that for developed countries it was technically and economically feasible to eliminate 90 per cent of methyl bromide, some OECD countries with a large agricultural sector, especially in Southern Europe and some American states, opposed it. The compromise reached in Vienna was a complete phase out by 2010.

Science and the public

The American media in the 1980s

In 1985, to all intents and purposes, the problem of the ozone layer did not exist for the American public; in 1986 there were over twenty articles in the national press, and in the following year there were more than twice as many. This ratio of growth demonstrates how suddenly attention to the subject changed due to the discovery and discussion of the ozone hole (Figure 6.2).

A breakdown of the press reports according to the various occasions that prompted coverage reveals an alternation between scientific and political occasions. Note that after the discovery of the ozone hole political occasions are more frequently the objects of coverage than scientific ones (Figure 6.3).

In this period, besides Rowland and Molina, the following actors came to particular prominence as speakers for the pro-regulatory alliance in the American national print media: Lee Thomas (EPA chief), Richard Benedick (chief

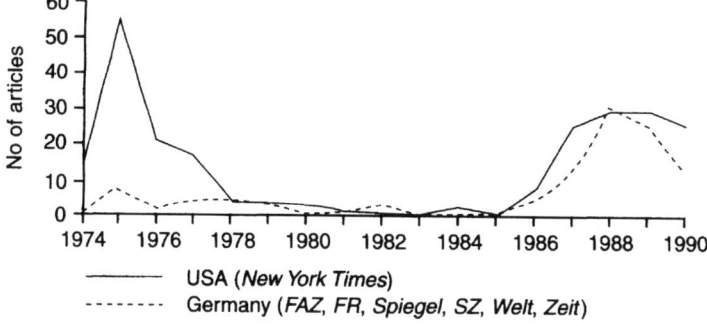

Figure 6.2 Media attention 1974–1990, USA and FR Germany

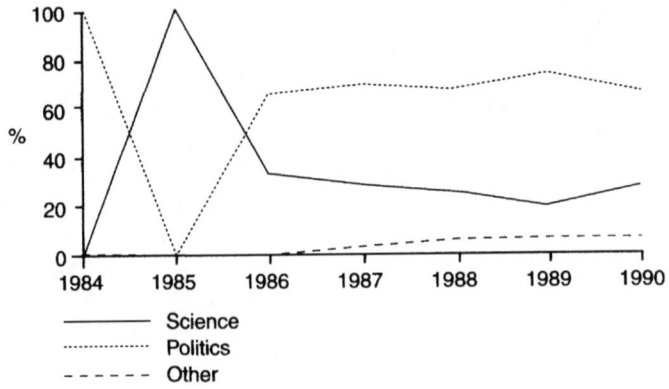

Figure 6.3 Occasions for US media reports, 1984–1990.

negotiator for the Montreal Protocol, External Affairs), Mostafa Tolba (UNEP director), Robert Watson (NASA), David Doniger (NRDC) and Michael Oppenheimer (Environmental Defense Fund). Other actors also expressed their concern, in particular regarding the Antarctic ozone hole.[13]

The phenomenal increase in attention given to the subject by the *Wall Street Journal* and the *Washington Post* is especially striking. All three major daily newspapers reported on the matter almost equally, although the *New York Times*, with over a hundred articles, remained in the lead. The peak period of attention in all three dailies occurs in 1989.

The German media in the 1980s

As in the USA, if on a smaller scale, media attention increased sharply from 1985 to 1988. The number of articles per year doubled every year from 1986 to 1988 (see Figure 6.2). The trigger for this increase in media attention in Germany was not the ozone hole, but rather the climate issue. From 1986, the press reported on the appeal made by the Energy Working Group of the DPG. The *Frankfurter Rundschau* (19 September) documented this appeal in full under the title, 'Through further warming the earth can become uninhabitable: The Deutsche Physikalische Gesellschaft warns of a climate catastrophe. Solar energy and nuclear power as possible way out.' *Die Welt* (15 January 1987) was somewhat less dramatic: 'Alarm signal for the worldwide climate: Scientists demand a quick ban on dangerous chlorofluorocarbons.' Table 6.1 provides an overview of the most important statements by scientists from German research establishments who demand urgent measures.

Just how long it took for the German press to turn the ozone layer into a political issue can be seen from the fact that the problem first appeared on the front page of a newspaper in July 1987 (*Stuttgarter Zeitung*, 27 July). The *Frankfurter Allgemeine* had a report on the front page on 21 March 1989, the *Frankfurter Rundschau* on 6 August 1990. *Die Zeit* was the first German print organ

Table 6.1 Press reports on demands of scientists, Federal Republic of Germany

January 1986	Declaration of the Energy Working Group (AKE) of the DPG
19 September 1986	Complete documentation of the declaration in the *Frankfurter Rundschau*
23 February 1987	German climate researchers: Take immediate measures
25 July 1987	Graßl; Ban on CFCs
11 July 1988	Crutzen: Take measures quickly
11 August 1988	Fabian: Ban on CFCs in spray bottles: Montreal Protocol insufficient
2 October 1989	Crutzen: Tighten regulations
1 December 1989	Zellner: As a precaution, a quick and efficient lowering of CFC emissions
6 August 1990	Bach: Cease production of CFCs

to mention the 'ozone hole' (18 July 1986); the *Stuttgarter Zeitung* the first to mention it in the title of an article on 4 July 1987, and *Der Spiegel* was the first to use the phrase on the cover. In 1988 there were six mentions in all of the ozone hole in article titles, and in 1989 there were five. On the cover of number 33/1986, *Der Spiegel* showed Cologne Cathedral standing partly underwater. The number bore the title: 'Ozone hole, melting of the poles, greenhouse effect: Researchers warn. The climate catastrophe.' In the following year there was a cover story on ozone depletion: 'Danger from a can: The ozone hole' (*Der Spiegel* 49/1987). What is striking is the high news value placed on science, which after 1989 was overtaken by politics (Figure 6.4).

The political career of the environment as a topic in the public consciousness was crucial for the political consideration of the subject. Media attention given to the ozone layer during the comparison period shows two high points, in 1975 and in 1988 to 1989 (see Figure 6.2). Already in the 1970s, there were newspapers in both countries that dealt with the problem in more depth than others (*New York Times, FAZ, Die Zeit*). After the discovery of the ozone hole, all the other papers caught up. What is conspicuous is the much more extensive coverage in the 1970s

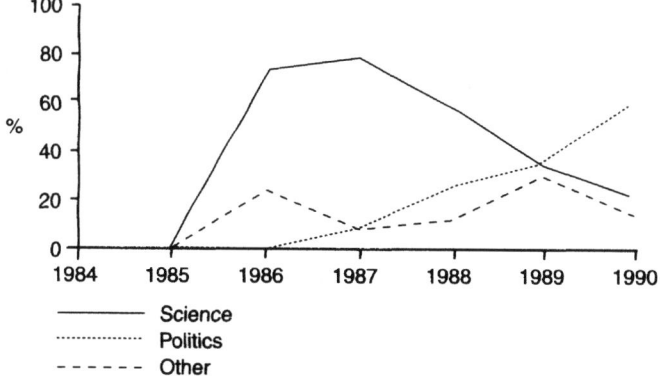

Figure 6.4 Occasions for German media reports, 1984–1990

in the USA. After 1978 there is an equally low parallel movement; the intensive media attention in the mid-1980s began somewhat earlier in the USA and lasted somewhat longer. In the USA there were researchers who went public as vociferous advocates. Such a group of actors does not appear in the Federal Republic until 1986.

While scientific publications usually do not receive a lot of media attention, this is different where scientific controversies are fought out (Goodell 1987). Conflicts, controversies and dangerous developments are more appealing to the public than day-to-day scientific practice. Sensations always sell. Various studies have shown that the media mostly stick to the facts when reporting on scientific issues but they distort scientific controversies. They over-represent scientists with a low academic reputation, especially when the topic is so salient that it is not only covered within the science section but moves to the political section. Since the media look for celebrity (and not reputation), they try to obtain statements from well-known scientists (Nobel Prize winners, scientific managers, academy presidents, etc.) and from representatives of opposing positions. Science journalists, for their part, form their own opinion and mostly present the information obtained by scientific speakers accurately. They usually do not question it by requesting additional sources. Goodell claims that they tend to vet and smooth out contradictory expert opinion.[14] He lists several reasons for the close relation between scientists and science journalists, among them the fact that science journalists are a small, highly specialised group who enjoy a high degree of autonomy within the editorial board since their bosses do not have the knowledge to judge the content of the reporting. Science journalists share the view of most scientists that the lay public is not knowledgeable enough to make judgements about scientific knowledge and practice. And, like the scientists, they tend to be highly enthusiastic about scientific progress.

Some scientists avoid the media, others are keen for it, some have to learn how to deal with the media.[15] The temptation is always great to exaggerate their own findings in order to get more funding. In risk debates, this may lead to over-dramatising. However, a 'cry wolf' strategy risks scientists losing their credibility by sounding the alarm too often or by timing the alarm in too obvious a way.

> As scientists we are divided – we have to warn but you cannot shout 'catastrophe' all the time. If you do that too often, in the end no one listens any more. But on the other hand: we alarm the public, well knowing that we do not know everything with certainty. You can overrate the problems, but you also can underrate them. The sudden appearance of the ozone hole is an example for the latter. A difficult situation for science.
>
> (Paul Crutzen in *Die Welt*, 23 October 1989)

Joe Farman who discovered the Antarctic ozone hole was asked why he did not publish his data earlier. He replied:

I sometimes feel we should have [acted sooner]. . . . On the other hand, the very fact that we delayed it until it was absolutely certain meant that there was never an argument. It was accepted. That's the real trouble with all these environmental problems. Too many people make too many noises all the time. Whereas if you only show it when you've got something to show, then, OK, people understand it.

(cit. in Roan 1989: 133)[16]

Over-dramatising can occur even unintentionally since scientists may not know at the time of making their statement whether it is justified, exaggerated or understated. Nevertheless, statements that are effective in public have to be spread by the mass media, which seems to add an additional element of over-dramatising.

If I wish to bring an issue like this to the attention of the public, it really has to be sensationalised, otherwise it won't be covered. And this happens all the way along. Either they are overly dramatic or can be interpreted that way, which then leads to all kinds of heavy duty publicity about it, which then leads to some disappointment: where are the dead bodies after all?

(USAD 22)

Advocates, like career-conscious scientists, likewise make use of dramatisation. It seems profitable to use such means to attract research funding or to establish or defend one's own field of study as a focal point of research. 'He who studies the fire, may not want to extinguish it' (Clausen and Dombrowsky 1984). This formula seems to express very well scientists' orientation towards self-interest. The conclusion could be drawn that atmospheric scientists are careful not to provide a solution to the problem, since they would then deprive themselves of their livelihood. Undoubtedly, this motive plays a partial role; the question is whether it is a dominant motive. As evidence to the contrary, besides the empirical observation that there are scientists who see themselves as speakers in the public interest, there is a theoretical objection. Other actors (above all competing research teams, the media, and potential backers who might be informed by the media) are also informed about this incentive structure, by which means a purely interest-based strategy could quickly undermine itself.

The public role played by scientific speakers can become problematic. In order to play this role, they must be visible and audible. Silent colleagues gaze enviously at the speakers standing in the spotlight and downplay the scientific value of their statements, because in their eyes such spokespeople talk to the press too much, vulgarising complex scientific connections with popular formulations; as Shils puts it, they wander from the centre to the fringe. On the one hand, such public engagement can help a career, particularly if one appears as the speaker for his discipline or his field of research and thereby gains his colleagues' approval. On the other hand, colleagues do not like to see someone bask in public glory instead of doing his daily work in the laboratory, in the lecture hall, or at his desk, reporting his research findings in the scientific literature. 'Anyone who has the time to talk to the media has no more time to do good science.' This, in many cases, will be the opinion of competing researchers in the field, especially of those who disagree with their prominent colleague's political interpretation of the scientific data. Jared Diamond (1997), writing on the career of Carl Sagan, has remarked that:

> scientists who do communicate well [to the public] are overwhelmingly at a senior stage in their careers. They wait until they have tenure and are thereby better able to withstand their colleagues' hostility. Young or nontenured scientists are relatively mute before the public because they realize that to be otherwise could mean the kiss of death.

The media select events to report according to their 'news value' (Schulz 1976). In terms of coverage of the environment, this favours a certain catastrophism or sensationalism (Brand 1995: 58). This is not, however, merely a strategy of increasing sales figures with dramatic events and panic mongering (de Haan 1995). Serious papers place great importance on objective coverage that, in part, goes so far as to expose reports that seem to amount to false alarms (Boventer 1993: 28). The *Frankfurter Allgemeine Zeitung* is a very good representative of this approach, and *Die Zeit* also sometimes follows this line. A speaker who disagrees with this approach describes it as follows:

> In the *FAZ*, environmental politics only appears when it can't be avoided, but then it's also certainly well founded. With *Die Zeit*, it depends on who the editor is; Herr Schuh could just as easily be with the *FAZ*, he's definitely competent, but he seems to work according to the principle of exposing the pollutant of the month as a bugbear.
>
> (GEAD 24)

The worse the potential danger appears in the public arena, to be sure, the less the politicians can afford to ignore the warnings. Politicians are then forced as a rule to act according to the imperative of 'blame avoidance' (Weaver 1986), funding further research and taking the results seriously. This does not mean, however, that they follow a precautionary course of action.

The construction of mass media attention (and media partisanship) is fascinating. It is striking how perfect the *timing* of alarming articles is. Some examples: concurrent with the opening of the London conference of parties to the Montreal Protocol (27–29 June 1990) a report appears in the international press to the effect that the ozone layer is thinning more quickly than expected, and that it could assume dimensions in the Northern hemisphere similar to those in Antarctica (*Financial Times*, 21 June 1990). Concurrent with the opening of the Copenhagen conference (23–25 November 1992), it is reported that the ozone layer is thinner than ever before: in the Northern hemisphere a decrease of 15 to 20 per cent has taken place. This report appears in November, after NASA had even predicted an Arctic ozone hole in early February (*Financial Times*, 14 November 1992).[17] Senator

Al Gore spoke of an 'ozone hole over Kennebunkport', the Bush family's holiday spot in Maine. The cover of *Time* read: 'Vanishing ozone: the danger moves closer to home.' The US Senate resolved immediately afterwards (in a vote of 96:0) to move the end of CFC production up from 2000 to 1995.[18] The forecasts of dramatic ozone depletion in the north did not come true. In April NASA reported that, due to a sudden warming of the Arctic air, the reduction of ozone had amounted to only 10 per cent. Last but not least: concurrent with the Vienna conference of the parties in November 1995 a new negative record is announced, this time for the size and duration of the Antarctic ozone hole. Three reports from the *FAZ* illustrate the staging of alarm. On 13 September 1995, the WMO reports that the rate of ozone loss is the fastest yet recorded since the early 1980s. On 8 November this estimate is revised: the WMO reports that the ozone hole has not grown further. On 1 December, two days before the conference begins in Vienna, German Environment Secretary Merkel calls upon the developing countries not to use up the transitional time period granted them, since the 'predictions of further enlargement of the ozone hole had been proven correct' (*FAZ*, 2 December 1995: 5).

7 Lessons

Alternative explanations

Policy networks are historical systems, whose state at any given time depends on their state in the past. Knowledge of a system's earlier states, however, is insufficient to predict further development; the latter is underdetermined by the system's earlier states. The ozone controversy of the 1970s influenced the ozone controversy of the 1980s, and the latter influenced the climate debate of the 1990s. No predictions can therefore be made; and so no attempt should be made, in reconstructing the historic data, to provide a teleological construction. Thus, in marked contrast to attempts to explain this case in a deterministic or reductionist way, my repeated reference to the contingencies of development.

Monocausal explanations?

The supposition made in Chapter 1, to the effect that in this case deterministic and reductionist explanations were offering too little, has been confirmed by reconstructing the case. No social system and no single logic of social action is dominant in cases like this. Instead we see combinations in which actors from various parts of society with various interests and world-views participate. In so doing we perceive self-reinforcing mechanisms (in particular through state subvention of research and media attention) that can be analysed more precisely with actor-centred instruments than by means of variables at the systemic macro-level. Here I summarise the objections to structural explanations.

The application of the network approach transfers attention from the contextual conditions of the process to the development of the network itself. Contextual conditions can be assumed to exist in the form of political, economic, public or scientific factors. Rival explanations based on these determinants are widely disseminated, not least because they offer monocausal explanations that are simpler than the network approach offered here.

The first explanation that one could conceive of is political. The argument would be that different political parties in power would lead to stricter or looser regulations. However, the record shows that the first regulations in the USA were passed under the Democratic administration of Carter, the second under Reagan's Republicans. In the USA, the distribution was across party lines; strong senatorial advocates included Republicans Stafford (New Hampshire), Chafee (Rhode Island), Durenberger (Minnesota) as well as Democrats Gore (Tennessee), Baucus (Montana), and Lieberman (Connecticut). In Germany the social–liberal coalition of Helmut Schmidt pursued a pro-industry policy; under Chancellor Kohl the government was dissociated from industry. Party political orientation (conservative/progressive) must therefore be dismissed as an explanation. However, one could assume that different political styles in different countries would lead to different regulatory outcomes.

Political style, which in the USA is held to be confrontational, while in Germany it is considered consensual (Moe and Caldwell 1994; Vogel 1986), certainly had an influence on the CFC controversy of the 1970s, since the USA dealt with industry fundamentally more harshly than the Federal Republic of Germany. The political style, however, transformed into its opposite in both countries in the 1980s. For this reason, political style can hardly function as an independent variable (see Table 7.1).

Turning to a final political variable, one could ask if the process at the international level was characterised by the hegemony of the USA, as the realist approach in the international relations literature would expect. Was it not American scientists, laboratories, satellites, managers and politicians who worked towards an international agreement? This is true, but cannot be attributed to the USA or a 'national interest' asserting itself. The dominance of the USA in the run-up to Montreal can be traced back to the growth of the advocacy network that had already begun in the 1970s. This dominance was only to be seen in the international negotiating process when the USA found itself on a pro-regulatory course and supported its actions by means of scientific resources. When the USA lost interest in the ozone question in the early 1980s and swung over to the anti-regulation camp, it could not be differentiated from other opponents of regulation. Since it was the advocacy network (and not the American government) that had been crucial to setting the American course, it would be more accurate to speak of the hegemony of the advocacy network, rather than of a hegemony of the USA.

In other words, the advocacy network conducted a struggle on several fronts, against the opponents of regulation both at home and abroad. Until shortly before the adoption of the official American negotiating position, the counter-alliance in the USA continued to attempt to hinder any international

Table 7.1 Political styles compared

Germany		USA	
1974ff	cooperative	1974ff	confrontational
1985ff	confrontational	1985ff	cooperative
1994	cooperative	1994	cooperative

agreement that contained more stringent CFC regulations than the Clean Air Act of 1977. And the role played by Richard Benedick as chief negotiator for the USA had certainly not been foreseen by the administration. After 1987 American scientists, politicians and bureaucrats worked towards tightening the Montreal Protocol. In so doing, they again worked together informally and mobilised both actors from other countries (at the government level and below) and a large number of symbolic resources.

Purely economic interests, too, can hardly explain the result. Above all they leave unexplained the fact that the Du Pont company came out vehemently against regulations in the first phase, yet became a leading proponent of regulations in the second phase. In Germany, Hoechst took a clear stance against regulations in *both* phases. It was hardly in the producers' interest to change over from production facilities that were neither amortised nor working to full capacity to new products or technologies. Moreover, a competition for alternative substances broke out in which firms from outside the chemical field took part. Their plans to find alternative substances for applications that had previously utilised CFCs were finally coming to fruition. In other words, as far as the CFC producers were concerned, the first priority was as far as possible to stay in business. The American manufacturers also had the economic interest of avoiding the negative effects of class action legal suits that for them was much more of a consideration than for their European or Japanese counterparts. For these reasons, Du Pont's change in attitude is a key event for the outcome of the controversy, and this in particular is in need of explanation. It is, to be sure, a myth that Du Pont changed its position because of any technological advantage.

Public opinion can lead to a change in governments' negotiating positions; governments, however, can also attempt to influence public opinion in their favour. When a nation publicly takes a specific negotiating position it also increases its bargaining power, if it thus constitutes a credible threat. Policy networks can produce a public expectation that a particular political position will be reached. Winfried Lang, in reference to the change in the European position at Montreal, poses the question whether this change is to be attributed to the coverage in the media, or rather to mechanisms of internal pressure:

> During the negotiations on the ozone layer it was the American delegation, which by means of continuous contacts with the media tried to build up a climate of public expectations which should induce still reluctant delegations (mainly those with EC-membership) to agree to substantial reductions of emissions. Further research will tell us, whether the relatively flexible stance finally adopted by the European Community was brought about by this manipulation of public opinion from the outside or rather by an internal process of rethinking threats and options.
>
> (Lang 1994: 175)

On the basis of the analysis presented here, it must be concluded that both the efficacy of the advocacy network in orchestrating public opinion and the pressure exerted in the course of the negotiations were important. The expectations raised by the media, as Lang describes them, may certainly have been much higher in the USA than in the EC countries. As shown in the last chapter (Figures 6.3 and 6.4), the USA and the Federal Republic of Germany show marked differences in terms of the occasions for press coverage after 1985. In American periodicals opportunities for coverage are of a political nature, while in German periodicals they are of a scientific nature. This indicates that in Germany there was a need to catch up with scientific information, while people in the USA were much more interested in the details of the negotiating process and the regulations. The absolute frequency of press articles was also clearly lower in Germany during the run-up for the Montreal Protocol. In the weeks before the Montreal Protocol was passed there were only two reports on the topic in German periodicals (in *Der Spiegel*, 17 August 1987 and in the *Frankfurter Rundschau*, 5 September 1987), compared to eight articles in the *New York Times* alone.[1] In terms of content, the German media largely follow the scientific interpretations.[2] Only late in the game do environmental groups integrate themselves to any extent into the advocacy network.[3] They take public action only when the problem is already well known and the world is alarmed. Their intervention cannot be seen as a key event.

Let us examine science as a possible independent variable. Was it not science that saw to it that ecological interests made it onto the agenda and took priority over economic considerations? Yes and no. Science accomplished this because it helped substantiate and legitimate regulations, in the course of influencing public opinion and political consultation. It was not 'science' (in the sense of a subsystem of society), however, but rather committed atmospheric scientists, who alarmed the public and the politicians and who were particularly active in the construction of a scientific-political transnational network. The mobilisation of corporate actors, material and symbolic resources was the clinching 'argument' that convinced the politicians that it was necessary to act, not least in the battle for votes.

Did the ozone hole as 'natural disaster', then, have an immediate effect on policy? Here, too, there is no unambiguous 'yes'. The ozone hole doubtless had this effect in the preparatory phase of the Montreal Protocol. It had a catalysing function, so to speak, working behind the scenes to shift the balance between opponents and advocates of regulation. In order for it to be able to play this role, it first had to be interpreted by advocacy scientists as a dramatic phenomenon. From rhetoric to visual presentation to application in political contexts, they managed eventually to utilise this 'sign from heaven' to good effect, turning it into a 'gift from heaven'. To this end, the scientists had to build up a sufficiently clear explanation of the processes in the Antarctic and of their own failure, without setting their credibility at risk (Zehr 1994). This process, in addition to leading to numerous scientific revolutions in the field of atmospheric science, gave the scientists a new understanding of their own activity. Many of them now believe in surprises and non-linear processes in nature – processes that they understand much too little to permit them to make reliable predictions.

Contextual conditions were thus not unimportant. They played a role by providing (sometimes surprising) resources for the networks and creating opportunity structures. Both, however, could only be made use of when there was a network in place that recognised the opportunities and utilised the resources. It is like a football game: a beautiful pass is of no use if there is no striker to reach for it. When this significant function of policy networks is overlooked, there is a tendency to see the various contextual conditions as causes. Various approaches do just this in a predictable fashion. As in the case of the explanation of the Antarctic ozone hole: every professional specialisation develops explanations in the area suggested by its own specialty. This is exactly what social scientists should expect. However, these partial explanations illuminate only certain fragments of the process. A more comprehensive picture comes into view once we manage to identify key actors and their goals and motivation, and to analyse the configurations in which they were to be found.

Two myths

In contrast to the approach advanced here, which investigates the workings of policy networks over time, many accounts of this case attempt to explain the success of the Montreal Protocols in a reductionist manner, by means of economic or cognitive factors. Such explanations seem attractive because of their meaningful simplicity and conceptual parsimony. Their beauty is flawed, however, because they are not too particular about the historical facts. Predominantly, there are two types of explanations. According to the first, the largest CFC producer (Du Pont) secretly researched substitutes and prepared to enter the market with them (thus winning a technological advantage that ultimately led to support for regulations). According to the second, scientists succeeded in explaining the linkages of cause and effect and communicating this to the political decision-makers, who consequently adopted stringent measures. Both identify important aspects of the process: on the one hand the role of Du Pont, on the other the role of scientists. Yet both need to be interpreted in a different manner.

The oligopoly thesis

Oye and Maxwell (1994) distinguish between two constellations that arise in the production of public goods and that are associated with the names of Stigler (1971) and Olson (1965). In the first constellation, regulations bring a benefit to only a few whereas the costs are spread widely. According to Stigler, firms are always interested in regulations that take the form of standard setting (in contrast to taxes) since this enables them to erect barriers of entry to new competitors and hence brings them extra profits. In the second constellation the benefits of regulation are enjoyed by many whilst the costs have to be borne by few. According to Olson, in this case the provision of the public good is unlikely, unless there is an incentive for the many to organise themselves (or if they can be forced to produce the good). Surprisingly, Oye and Maxwell interpret the ozone case as an instance of a 'Stigler-constellation'. As will be shown, this view relies on a questionable interpretation of historical facts, since *barriers to entry* played a minor role in this case.

Back in the 1970s, CFC producers tried to find appropriate substitutes for CFCs and make extra profits in a 'high-tech, high-risk' market (Maxwell and Weiner 1993). But it did not work out: the replacement of CFCs as propellants in spray-cans was done by old ('low') technology (for example, propane), thus diminishing the value creation. However, Oye and Maxwell go on to claim that the Montreal Protocol can be explained by the drive to make extra-profits in a high-tech market. The weakness of this approach lies in the fact that it cannot answer two crucial questions: first, why did CFC producers stop the research into CFC alternatives in 1980 but took it up again in 1986? And second, why did Du Pont take the first step to get out of CFCs at a time when European producers were opposing it? There must be other factors involved. This makes a parsimonious explanation of the Oye/Maxwell/Stigler type less convincing.

Also arguing from a rational choice perspective, Sprinz and Vaahtoranta (1994) hold that the trade-off between economic costs and ecological vulnerability determined the different national positions in the international negotiation process. Based on their own estimates and calculations, the authors argue that in the negotiation process a group of countries dominated for which the potential damage was greater than the costs incurred if they switched to more benign products. As indicators they take the estimated future costs of treating skin cancer. This approach assumes a rationality on the part of the actors that in reality hardly played a big role. For one, the depletion of the ozone layer is a global phenomenon, its effects are not limited to increased incidence of skin cancer, and regional variations as an effect of increased UV-B radiation are not established today, hence they were not known in the 1980s (cf. UNEP 1994). In fact, the ozone layer over the tropical countries seems to have remained stable – which could have led these countries to veto any international efforts to reach an agreement. But even tropical countries could not rule out the possibility that further chlorine emissions into the atmosphere would change the global atmosphere in a way which would imply disastrous consequences for tropical countries too. In fact, in the meantime, nearly all nations – including tropical ones – have signed the Montreal Protocol.

However, the rational choice approach of Sprinz and Vaahtoranta fails for another reason. Based on their model assumptions about economic cost and ecological vulnerability they classify the main protagonists of the international negotiations into four categories: bystanders, pushers, draggers and intermediates. The result is that two central actors (the USA and Germany) behave differently in reality than the model would predict. Both countries were pushing for progressive measures (from around 1986) but were expected to be 'draggers' (Germany) and 'intermediates' (USA) respectively. Confronted with this anomaly, the authors postulate technological developments that have allegedly taken place during the period of negotiations and opened new opportunities for both countries.

There is poor evidence that there was one producer of CFCs (Du Pont) that had achieved a technological advantage over its competitors through secret research on possible substitutes for CFCs – although this is the most pervasive and most widespread myth surrounding the Montreal Protocol. Ironically, Sprinz and Vaahtoranta vindicate this myth by referring to a source who actually denies it. Benedick, whom they quote in their support, writes:

> Some Europeans suspected that the ... U.S. companies ... had endorsed CFC controls in order to enter the profitable EC export markets with substitute products that they had secretly developed. This suspicion was unfounded. ... Events after the Montreal Protocol conclusively demonstrated that there had been no secret substitutes on the shelf.
>
> (Benedick 1991: 33)

Apart from Benedick, the authors adduce two more sources (Morrisette 1989; Sebenius 1992) – neither of whom support the thesis of a technological advantage. Sebenius relies on Benedick and gives no further evidence. He simply claims that Du Pont was ahead of its competitors (Sebenius 1992: 358). Morrisette repeats Du Pont's arguments that substitutes would become only profitable if and when government regulations set the right incentives. This argument actually reverses the causality implied in the mythical account of the technological advantage (Morrisette 1989: 816).

As noted earlier, the figures of Du Pont's investment in alternative substances are telling: the firm invested 5 million US dollars in 1986, 10 million in 1987 and 30 million in 1988 (Reinhardt 1989). After 1987, the yearly average was ten times the pre-1985 level. Competition for substitutes began around 1987 and intensified in the years to come. Firms were facing the following main problems:

- substitutes were unlikely to cover 100 per cent of the prior use, i.e. there were competing substances and technologies kicking in 'from outside';
- the alternative substances were subject to time-consuming toxicological tests prior to their admission;
- firms had to find production processes for the alternatives.

The first point indicates that the primary goal for CFC producing firms was to stay in business as long as possible. Contrary to Oye and Maxwell's claim, this turbulent market did not allow the established chemical firms to defend their position, let alone to erect barriers against newcomers. The second and third points indicate that a time gap existed between the testing period and the mass market. In this period firms were building pilot facilities, partly by means of common research networks.

Scientific consensus?

Peter Haas has put forward a much acclaimed model to explain international policy coordination. Against neo-realist and neo-liberal approaches in international relations literature, he rightly maintains that under conditions of uncertainty epistemic communities play a decisive role. Haas (1992a: 17–18) holds that members of epistemic communities gain access to the political system by virtue of their professional training, prestige, and reputation for expertise in an area highly valued by society or elite decision makers. Applied to the CFC case, he states that the epistemic community consisted of a

> knowledge-based network of specialists who share[d] beliefs in cause-and-effect relations, validity tests, and underlying principled values and pursue[d] common policy goals. Their orientation is perhaps best expressed in the words of one member, who voiced his willingness to accept the 'plausibility of a causal link without certainty'.

> (Haas 1992b: 187–8).

Still, several problems remain with his approach. If we believe him, it was the common scientific understanding of the case that led policy makers to adopt stringent measures.[8] However, in the two main research areas there was no established scientific knowledge at the time (1987). These were the area of global ozone trends and the area of polar ozone ('ozone hole'). In the first area the question was whether there existed a significant downward trend in ozone concentrations around the globe. In the second area scientists tried to understand the unexpected, dramatically low ozone concentrations over springtime Antarctica. The first issue was mainly about establishing an observation ('do we have a downward trend or not?') whereas the second started from an established observation ('yes, there are large losses of ozone over Antarctica') and tried to explain it.

The causes of global ozone destruction (in middle and northern latitudes) were not well understood then (and are probably not today). Various theories existed but there was no consensus. If one looks at the issue of polar ozone, it is clear that in 1987 there were still many theories competing to explain the ozone hole, several of which involved CFCs, and two of which claimed that the phenomenon was caused naturally (Cagin and Dray 1993; Nance 1991; Roan 1989; Shell 1987). In 1988 – after the signing of the Montreal Protocol – a consensus emerged that still was limited. The Ozone Trends Panel established global ozone losses in 1988 as a scientific fact, without being able to give the reasons. At that time only the hole over the Antarctic could be explained fairly well.[9] While Haas recognises the problem of the international cooperation of the Montreal Protocol being concluded one year before this 'consensus', he does not adequately account for it.[10]

Why was cooperation possible in 1987 even though the (partial) scientific consensus did not come about until one year later? It seems as if Haas goes astray because he makes two questionable assumptions: that policy makers turn to experts to ameliorate uncertainties, and that they act on the basis of the consensus view held by these experts. In so doing, Haas overlooks two other possibilities that might be more effective in a process of policy making under uncertainty. First, advocacy scientists might appear on the scene even before decision-makers realise that there are problems that involve uncertainties. Because of their political motivation, such scientists might want to try to influence policy options from the outset. Second, the reliance on expert knowledge might be more limited than Haas perceives. What we frequently observe is that the specialists are divided and that political decision-makers instrumentalise one position for their purposes. After all, politics is not based on science, but on political judgement. This should makes us think again about the role of scientific arguments in the policy process.

At first sight, the account provided by Litfin (1994) comes close to my reading. She emphasises the point that proponents of regulations – far from having reached a scientific understanding of the case – were contextualising pieces of information in a skilful way. It was not objective knowledge but powerful interpretations that made the difference. Nonetheless, on two counts we seem to disagree. The first disagreement pertains to the role scientists played in the process. I argue that their contribution to the final success was large, even in terms of policy recommendations. Litfin thinks that it was primarily science (namely, information about facts) that drove the process. She notes that 'scientists were important actors in the process, but saying the issue was science-driven does not say that the scientists themselves were the driving force . . . they rarely made policy recommendations . . . other contextual factors determined how [science] would influence policy' (Litfin 1994: 115). My interpretation of the material establishes that advocate scientists who were in favour of regulations played quite an active role. My second disagreement with her is about the description of the 20-year process. She depicts it as the discourse of precautionary action moving from a subordinate to a dominant position. While this holds true for the post-1984 period, it does not describe the whole process. The precautionary discourse had become dominant at the very beginning of the controversy. It suffered a drawback in the early 1980s when the Reagan administration took over. But soon it revitalised and expanded.

The first period of controversy in the USA had the result that legislation on the ozone layer was largely based on the precautionary principle. The Clean Air Act of 1977 authorised the administrator of the Environmental Protection Agency, EPA, to regulate 'any substance . . . which in his judgment may reasonably be anticipated to affect the stratosphere, especially ozone in the stratosphere, if such effect may reasonably be anticipated to endanger public health or welfare' (Clean Air Act, cit. in Benedick 1991: 23). The key here was that no conclusive proof was necessary for action to be taken – just a reasonable expectation (EPA 1987). Nearly a decade later, this concept shaped the American position on international controls (Betsill and Pielke 1998).

Lessons for other global environmental problems

The learning capacity of modern societies

The results of this study raise several general questions, only three of which will be taken up here. First, the question of whether modern societies are capable of avoiding catastrophic developments arising from their impact on the natural environment. This touches on the problem of the predictability of global ecological problems. Second, the question of where in modern societies the knowledge that could provide information on such problems can be found. Here there is evidence of a specific degree of power held by natural scientists, which until now has clearly been greater than that of social scientists. The third question is whether diffuse interests can be organised, and whether global common pool resources can be conserved.

It is impossible to classify global ecological problems either as generally predictable or as generally unpredictable. Two prominent global problems, the greenhouse effect and population growth, were already under discussion more than a hundred years ago. The problem of possible damage to the ozone layer was first recognised a good twenty years ago; initially aeroplane and rocket emissions were presumed to be the possible causes, later CFCs – a material that until then had been considered totally non-polluting and with no side-effects for the environment. This prompts fears that among the thousands of industrially produced chemicals, there may slumber many other potential catastrophes.

> If someone wants to kill a new bug, they will find a formula and it is quite easy to make and then you have a substance which has never been in the world before. And it can be on a mass market in amounts which matter. You can flood the world within 10 short years. There is none of us clever enough to take this formula and look at it and say 'It will have this effect on the world'.
>
> (UKAS 44)

This scientist considers it to be pure good fortune that only bromine and chlorine, but not fluorine, have dramatic effects on the environment:

> It is not a small change we've made with chlorine. It is seven times since the 1930s. Luckily fluorine did not matter, it is just sheer luck. So you have to be greatly pessimistic that technologically we are extremely clever and environmentally we are extremely stupid.
>
> (UKAS 44)

Paul Crutzen, too, has the impression that humanity stumbled within a hair's breadth of catastrophe:

> Bromine [is] almost a hundred times more dangerous for ozone than chlorine on an atom to atom basis. This brings up the nightmarish thought that if the

chemical industry had developed organobromine compounds instead of CFCs – or alternatively, if chlorine chemistry had behaved more like that of bromine – then without any preparedness, we would have been faced with a catastrophic ozone hole everywhere and at all seasons during the 1970s, probably before the atmospheric chemists had developed the necessary knowledge to identify the problem and the appropriate techniques for the necessary critical measurements. Noting that nobody had worried about the atmospheric consequences of the release of Cl or Br before 1974, I can only conclude that we have been extremely lucky, which shows that we should always be on our guard for the potential consequences of the release of new products into the environment. Continued surveillance of the composition of the stratosphere, therefore, remains a matter of high priority for many years ahead.

(Crutzen 1996: 1771)

If we should be on our guard, but we have no certain knowledge, how then can we recognise untoward global developments in time? Delphi surveys, for example, have determined what, in the opinion of scientists, are the most urgent global problems (cf. Stewart 1987; Wilenius 1996). In the matter of estimating individual uncertainties, risks and potential deleterious effects, however, no consensus can be expected from the experts. This is a consequence on the one hand of the role world-views play in scientific practice, and on the other of disciplinary specialisation. To some extent the two processes intertwine, as, for example, when one speciality makes findings that exonerate industry, while another produces incriminating results. It is thus an illusion to suppose the protection of scientific autonomy would be sufficient to produce the right kind of knowledge. The technocratic policy model is thus rendered obsolete. In its place, a new model emerges: one that allows, and requires, that scientific controversies take the form of public contention. As a further result, the bringing together of scientific expertise from different specialities or disciplines can reveal a wider spectrum of potentially serious problems, without advancing at the same time to understanding or explanation. The politicisation of science can have a catalytic effect on this process. But this could also mean that greater demands are made on successful scientists: they must move beyond the horizon of their own specialisation and that of 'pure science'.

Catalytic processes in the atmosphere and in society

The knowledge that trace gases are responsible for the ozone balance in the atmosphere is fairly new. The concentration of these gases amounts to millionths of a part per volume of air. The catalytic mechanisms that can repair the ozone layer in the long term were released down on Earth, within society, by a tiny group of actors. Before this could happen, two itineraries had to cross: the route of the CFCs on their way into the stratosphere and the route of the atmospheric scientists on their way back to Earth from investigating other planets. For the stratosphere had gone unnoticed for a long time; even at the beginning of the 1970s it was a scientific no-man's land. Meteorologists busied themselves with the troposphere, the region where weather happens. The preferred research area of the aeronomists was predetermined by the financial support of the US Department of Defense, and consisted of researching the re-entry of rockets into the atmosphere – a process that occurs in the mesosphere (above the stratosphere). And after NASA took over the lion's share of funding research in the field of aeronomy, the atmosphere of other planets took priority ('Is there life on Mars?').

The atmospheric scientists returned to Earth at about the same time as the 'blue planet' entered public consciousness, where it was seen as something 'tiny, delicate, fragile' (Carl Sagan) in utter contrast to Mars, Venus and the moon, which seem to harbour no life:

> Discovery of the earth took place during the journey to the moon. When in July 1969 Neil Armstrong uncoupled himself from spaceship Apollo 11 and touched down in his landing-craft on our neighbouring planet, he found only barrenness, emptiness, and icy silence—but when he looked backwards, he went into raptures. How different the earth appeared! Shimmering blue, it floated like a spherical jewel in pitch-black space. . . . Amid the desolate expanses of the universe the old earth reveals itself to be the inhabitable, the absolutely special, star that is our home.
>
> (Sachs 1999: 110)

Meanwhile, the vulnerability of Planet Earth has become a topic of discussion in the world public arena – due to global environmental problems. This is highlighted by the CFC problem. When the creeping threat had turned into an acute threat following the discovery of the ozone hole, the international community felt compelled to cooperate. Environmentally committed scientists played an important socio-political and catalytic role in these events. However, as we have seen in this study, before catalytic processes can begin, a critical mass of reactive 'substances' must be present. Institutional factors had ensured that the critical mass in the USA was greater, and developed earlier, than in the Federal Republic of Germany. Here a process of 'catching-up' took place: in a short time the public became alarmed, scientists were activated, and politicians correspondingly grew more ready to take action.

Scientists who have a high public profile do not fall from the sky. Many are faced with the problem of reconciling the role they actually play with the social expectations of them (the 'ideal of science'). Because too great a public political commitment can damage their scientific credibility, they set great store by consensus on purely scientific questions. In particular, they emphasise that without scientific findings, they would not have taken sides. Not all scientists, however, are convinced by these findings at any given time. Moreover, their partisanship also varies according to their professional specialisation and personality. Their strategies of purification are the necessary complement to a process of hybridisation in which they rely on their practical competence of judgement.

Hybridisation

Will it be possible, despite increasing differentiation within modern society, to achieve the integration of disparate practical skills?[11] What institutional arrangements are suitable for this? It seems that two problems have to be solved on the way: first, the broadening of the (specialised) knowledge base; second, acquiring the competence to make judgements based on investigations that produce ambiguous findings. While the broadening of the knowledge base can be accomplished institutionally by means of multi- and interdisciplinary research projects, as yet there is no institutional solution in sight for the second problem. On various occasions it has been proposed that the model of the medical profession should be followed, according to which scientists are educated to acquire informal rules for decision-making in the course of their practical work. Their competence would then originate in a combination of theory and practice (Böhme and Schramm 1985; Marcus 1988).

In the research field of the ozone layer, certainly, a fruitful and successful collaboration between atmospheric chemists and dynamic meteorologists has developed; this collaboration, however, has not spread to other fields – to biology, for example. Can we learn something from this successful cooperation that can be applied to other cases? In my view, it is above all committed and motivated scientists who have managed to develop a more comprehensive perspective in this manner. The trailblazing studies in atmospheric science were almost all produced by outsiders. Rowland, Crutzen and others are a perfect example of this. Rowland was a newcomer to the field of atmospheric chemistry when the Molina–Rowland hypothesis was published. At the time he said of himself: 'I'm a well-known unknown' (cit. in Roan 1989: 5). In the course of the controversy he dabbled with great success in an additional area of expertise (in statistics, in the context of the OTP studies). Without his motivation it would probably have been difficult to turn the actual measured ozone depletion so quickly into a scientific fact.

> The whole issue would not have developed to this point if Rowland hadn't been so missionary about it. If it would have been treated objectively, scientifically, as I would have liked it to be done, it probably would never have been treated as a serious issue by the public and by politicians. If he hadn't stirred up the Greens and the politicians. . . . He must have spent an enormous amount of his time and effort going around lecturing, talking. He really barnstormed. He went to every little town and every little community, delivering his speech. I thought this isn't the way to do science, but I think he was probably right, because he believed in it.
>
> (Interview with Jim Lovelock)

Crutzen was an autodidact; Cicerone, Molina and Stolarski were young and not yet established atmospheric scientists.

Politically committed and publicly visible scientists with a broader knowledge base and a practical sense of judgement do not grow on trees; they become like this only after a long process of scientific and public commitment. This is a thorny path, particularly considering the ambivalent incentive structure in terms of public commitment. It would be wishful thinking to believe that the increasing magnitude of environmental risks automatically increases the willingness of social actors (in this case among scientists) to take risks, thereby creating a balance in accordance with the Hölderlin–Marx dictum: 'Where there is danger, the means of rescue also grows', or 'Mankind only sets itself problems that it can solve'.

The developments described above seem hardly reconcilable with two older approaches in the sociology and philosophy of science. First, scientific practice as described here does not conform to the normative structure of science as Merton conceived it. Not only do the self-seeking ('unethical') cases of (unsanctioned) violation of norms speak against it (see Chapter 3), but above all the almost universal breach of the precept of organised scepticism and impartiality.[12] Scientists who are convinced of a case, however, even when they cannot prove it, certainly possess many times more motivation and persistence than others in seeking funding and attention[13] – assuming that they do not lose contact with the core of the research group, and continue to differentiate themselves from interest groups. This could become a general pattern for scientific practice under conditions of greater competition. Second, in the form of scientific practice presented here, Popper's criterion of falsification is not really observed. In the course of the long controversy, both sides are charged with data that contradict their theories, prompting reactions of defence or attack that go so far as to alter the formulation of the problem – for followers of Popper a typical sign of ideologisation and immunisation of one's own theory (and so of unscientific behaviour). Such problems do not beset the more recent approaches to the sociology of science, because these approaches proceed from premises of the interpretative flexibility of scientific findings and of the importance of social factors in creating new knowledge. The findings presented here confirm this. This certainly does not mean that any scientific finding can lay claim to the same validity. The limitations to validity that arise, however, are related above all to the temporal and social dynamics of the development of knowledge (Adam 2000). In the course of a controversy, points that have been 'checked off' are no longer pursued; those who nonetheless insist on pursuing them are marginalised. Once a core group of researchers has agreed upon a testable set of hypotheses in this manner, it can then be confirmed or falsified analytically and experimentally. This occurred in the case of the ozone hole by means of standardisation and field experiments, which to some degree attained the status of *experimenta crucis*.

Practical knowledge

Uncertainty, to be sure, does not mean complete ignorance; however, they may be related. Whenever at least two existing theories are taken seriously by the scientific community, we find ourselves in a condition of uncertainty, since we cannot say which of the two is correct. Elster (1979: 384–5) goes one step further: 'The larger the number of competing theories, the larger the probability that they are all false'. The problem becomes more acute if uncertainty and irreversibility coincide, because under such circumstances it is impossible to generate enough certain knowledge before it is too late to turn back. Critical threshold values can only be identified *post festum* or, as William Blake put it: 'You never know what is enough unless you know what is more than enough' (*Proverbs of Hell*). When there is no generally accepted theory, we can turn to practical competence. Elster is sceptical of this possibility, since we do not know who has this competence and who does not. He sees competence in successful politicians and entrepreneurs (because they have 'survived'), rather than in scientists and administrators who normally occupy secure positions in protected niches. In particular, he thinks that we cannot assume that scientists provide informed and undistorted judgements, since they adhere to one of the competing theories. This may be true; however, this seems to ignore the role of advocacy scientists as socio-political actors, who possess practical competence and judgement that is constantly subject to tests of credibility in a public controversy. The better they pass these tests in the course of time (compared to competing actors), the more trustworthy they become.

But what are the chances of finding practical knowledge (Stehr 1992) outside the natural sciences, yet still within the realm of science? Lepenies (1989) sees sociology of science, history of science and theory of science as possible 'secondary orientational disciplines'. Dismissing alternative research ethics, the acquisition and distribution of orientation knowledge take the highest priority: 'It's a question of developing a new scientific mentality. We're faced with a problem of socialisation' (Lepenies 1989: 155). The findings of my study suggest the conclusion that there is a country-specific difference in this regard, as shown by the fact that in many cases American scientific experts, who do more of their work at the edges of science and in public, provide such practical knowledge. If my analysis is correct, Lepenies' suggestion that sociology and science studies perform a special function is certainly not invalidated. In addition to the tasks he mentions (destroying the three myths of science's autonomy, of cumulative increase of knowledge and of Western rationality), the possible socio-political role of scientists must also be considered.

But all this still lies ahead of us. Before sociology can even begin to come to grips with questions of decision under uncertainty and the representation of diffuse interests, the natural scientists have set the route – and not always in the direction of increased risks.[14] Atmospheric scientists even reproach social scientists for not having done their homework and sketch their own proposals for solutions (cf. Hasselmann 1998). When social scientists attempt to react, they are often overpowered in the face of natural science-based explanations. Sometimes they tend towards cynicism or moralism and evaluate findings or interpretations selectively to substantiate their own (involved or detached) position. Beyond this accommodation of scientific findings to their own ideology, analyses based in the sociology of science could describe the strategic and rhetorical practices of scientists in order to understand the process of producing facts and institutions. Sociologists, however, often stand to one side in a manner that relativises the risks, referring to mass-media cycles of attention (Downs 1972) and to the likely disappearance of the problem. Thus the questions are taken less seriously than they seem to merit in the current situation (Wildavsky 1994). In fields where they are taken seriously, this conviction (often dressed up in moralistic terms) induces scientists to endorse specific political demands. The alternative to relativistic cynicism or moralistic concern is an analysis of the discourses of the natural sciences that will form a clear picture of the relevant interests and ideas of central actors in such controversies (examples include Hannigan 1995; Jasanoff and Wynne 1998; MacNaghten and Urry 1998; Yearley 1996).

The common thread binding esoteric science to mundane politics, the Antarctic sky to chemical factories, the global threat to the next election and to the United Nations (Latour 1993), was traced by politically committed scientists and led to numerous complications in the historical process. Any serious analysis of these complications has to tie together the development of knowledge claims and institutional structures, of scientific evidence and world-views of scientists, international cooperation and national political events. The present work represents such an attempt to analyse the 'complex chemistry of international ozone regulations' (Parson and Greene 1995). The findings presented here await investigation in further empirical studies.

Five theses

The main focus of this book has largely refrained from investigating the correctness of scientific statements about the threat to the ozone layer or the adequacy of political decisions in this regard. The temptation to derive possible lessons extending beyond the concrete case in question, however, should at least partially be given way to. To this end I outline five theses.

Thesis 1: the representation of diffuse interests in preserving a common pool resource demands speakers

The preservation of a global common pool resource requires placing the problem on the international political agenda. This can occur through international organisations or as the result of the commitment of individual nations. As long as there is no incentive to voluntary changes in production, either through cost-effective technology or through public pressure, the most rational course of action for the producers is to maintain production. The preservation of a common pool resource thus requires speakers or public interest groups who take on the representation of diffuse interests. This occurs first in national contexts, where the corresponding political options are institutionally anchored.

Thesis 2: transnational relationships are crucial

Once domestic political developments lead a nation to take up a position in favour of precautionary measures, it will influence obstructive and undecided states on the international level. Beside this well-known form of intergovernmental relationship, transnational relationships are becoming more and more important. Transnational political networks comprise actors from various backgrounds and hierarchical levels, governmental representatives and non-governmental organisations. Here it is necessary to emphasise the relative autonomy of representative corporate actors. These actors often act in the name of their principals (governments, firms, organisations) without complete authorisation to do so. This permits them to create new facts that are acknowledged after the event.

Thesis 3: scientific laboratories provide important resources for decisions under uncertainty

By tying themselves to the results of scientific research, actors can attempt to substantiate, 'rationally' to a certain degree, their choices for acting in public discourse. The prerequisites for this substantiation, to be sure, are laboratories, which is why only nations with sufficient scientific infrastructure play a role in international controversies dealing with decisions under uncertainty.[15] Like the licence to print money and the monopoly on power, laboratory data represent a means to power that can also be used in the political process. The potential for legitimation exists for politics above all when, despite the existence of scientific uncertainties, a link to political options can be forged. This requires that the public advocate such a policy. If a group of scientists succeeds in alarming the public and appearing as speakers for the interests of those affected, a probable self-sustaining dynamic develops among science, politics and the public. Industry, in such constellations, loses some of its potential influence.

Thesis 4: when scientists appear as speakers for diffuse interests, they play a special role in the contest for public credibility

Public opinion and the orientation of the media can exert pressure on firms and politicians to act; the choice of themes and the terms of discussion, however, are often made by experts who sympathise with a coalition of advocates or even act as its public speaker. The more credible one side appears in public, the greater the likelihood of strengthening that side and eventually gaining hegemony. Alarm signals and crises are crucially significant in this process.

Thesis 5: in long-lasting socio-political controversies, the transfer of opinion leaders from one camp to the other causes a domino effect

In antagonistically structured political areas where two camps face off for years in a protracted struggle for hegemony, one network can win allies and/or resources from the opposing network by an attack on the opposition. The growth of one side at the expense of the other may even produce a domino effect and thus a lasting and decisive change in the balance of power between the networks. Whenever one side succeeds in removing resources or allies from the opposing alliance and recruiting them for itself, the inherently dynamic process turns into a chain reaction, above all if a large number of actors outside of both networks remains undecided for a long period and then suddenly becomes active. This is particularly likely whenever opinion leaders move from one camp into the other. Opinion leaders are actors in relation to whom, at any given time, several other actors orientate themselves.

8 Epilogue: the example of climate

The debate over climate initially found itself in the lee of the ozone controversy. Important actors in the ozone debate are also active in the climate debate. Above all, they drew three lessons from the ozone case that they applied to the negotiation process regarding climate: standardisation of scientific results, technical indicators for the solution of the problem, and the institutional separation of the framework agreement from the catalogue of measures taken (protocol).

Although the problem was recognised long before the ozone layer itself (Arrhenius put forward the 'greenhouse theory' almost a hundred years ago and calculated the rise in average temperatures by a doubling of the concentration of CO_2), it did not arrive on the political agenda until the 1980s. Up until then, even scientists dismissed hypotheses of anthropogenic climatic change as a fringe opinion (Hart and Victor 1993). This changed after 1985, with several international conferences in the course of which the topic established its credibility. The discovery of the Antarctic ozone hole around 1986 and the extreme heatwave in the USA in the summer of 1988 contributed to this process. In the same year, the World Conference on Atmospheric Change took place in Toronto (Canada), where extremely ambitious goals were formulated (goals that were never again to be repeated, such as 20 per cent CO_2 reduction by 2005, based on 1988 levels; cf. Bodansky 1994). It was also in 1988 that the international community through UNEP and WMO formed the Intergovernmental Panel on Climate Change (IPCC). This committee was supposed to alert the political decision-makers to scientific findings on climate change. Earlier scientific reports, particularly those from OECD nations, were met with distrust on the part of several developing nations. The warnings about climate change were regarded less as robust scientific findings than as an expression of environmental activism. The establishment of the IPCC as an institution of the United Nations can be seen as a deliberate attempt to take the wind out of the sails of such criticisms.

The central findings of the IPCC are as follows (Houghton et al. 1996). The global average surface temperature has increased in the last hundred years by 0.3–0.6 degrees Celsius. The global temperatures of recent years are among the warmest in history. We are living in one of the warmest periods of the last six centuries. The warming is not uniform, however; some regions have even become cooler. Global warming thus means primarily climate change; change in amount and patterns of precipitation (more precipitation in the vicinity of the poles, less in the already arid equatorial regions), rise in sea-level, disequilibrium in plant growth (more weeds relative to useful plants), spread of infectious diseases, and so on. Although the signal of anthropogenic influence is still obscured by the noise of natural variability, the assumption is that current climate changes are connected with human activities, in particular with emissions of carbon dioxide (64 per cent), methane (19 per cent), nitrogen oxides (6 per cent) and hydrocarbon compounds (11 per cent). Computer simulations of the climate system calculate an increase in surface temperatures of 1–3.5 degrees Celsius over the next 100 years. Pre-industrial CO_2 concentrations in the atmosphere amounted to approximately 280 ppmv, nowadays have reached 368 ppmv, and in the next 50 years may reach twice the pre-industrial concentration. In order to achieve stabilisation at the present level, these emissions would have to be reduced by more than 60 per cent (Wuebbles and Rosenberg 1998).

Ozone and climate: differences and commonalities

Compared to the ozone case, the IPCC resembles the model of an epistemic community much more: it was designed to forge a consensus in the field of climate policy.[1] The scientific working group of the IPCC (Working Group 1) can certainly point to some successes in this regard.[2] This is hardly surprising, considering that the scientists represented in the IPCC were nominated by government representatives who were themselves concerned about global climate change (O'Riordan *et al.* 1998: 369). The result is an 'orchestration of consensus' (Elzinga 1995). This orchestration followed from the experience in the ozone example. In that case, the existence of several partly contradictory scientific assessments had served to justify a 'wait-and-see' attitude. It was not before 1985 that a single international report was put together. As a result, the climate debate was intended from the beginning to lead to scientific consensus, if only minimal. This consensus was in essence achieved, but at a cost: scientists with dissenting opinions are not represented in the IPCC. They disseminate their position vociferously, however, through the media.

By institutionalising international scientific assessments, the architects of the IPCC drew an essential lesson from the case of the ozone hole. They tried to reach a consensus view on the scientific aspects of global climate changes, thus forming an epistemic community (Haas 1992a). Founded in November 1988, the IPCC initially sailed in the waves of enthusiasm created by the successful Montreal Protocol. Its role is to review and assess the published scientific literature on climate change, its costs, impacts and possible policy responses. It also plays a role in assessing scientific and technical issues for the UN Framework Convention on Climate Change (FCCC) (Shackley 1997). Therefore, the IPCC is modelled precisely after the WMO/UNEP assessment reports in the ozone case. In both cases, a standardisation of scientific knowledge is seen as instrumental to get the right policy decisions. This follows a linear or 'technocratic' policy model according to which first a scientific consensus has to be reached which is then transformed into political decisions.

It has been rightly remarked that insofar as scientists adhere to this view, they must be regarded as rather naive (Shackley and Skodvin 1995). Others have argued that the IPCC has primarily served the self-interest of the participating scientists in that they attracted huge funding resources and therefore stayed away from advocating specific policies (Boehmer-Christiansen 1994a, 1994b). To this has come the reply that the avoidance of policy advocacy in IPCC reports is rooted in a desire to make the scientific information as effective as possible: 'For scientific information to be believed by the majority of participants in policy debates, it must be even-handed and not favour particular political or economic interests' (Moss 1995). Both views seem to accept the linear model of a science input that is transformed into a policy outcome. Without doubt, the IPCC has succeeded in establishing a shared understanding of climate change that is accepted by many participants involved in building the climate change convention. But did it achieve enough?

From the consensus assessments of the IPCC, rough goals could be derived; though not, however, solutions of the conflict between countries and groups of countries. But perhaps this is expecting too much from the scientists, given that they have resisted making detailed recommendations to the politicians. It is noteworthy that the FCCC has taken over another element of the Montreal Protocol role model: it ensures that tougher control measures are possible if new scientific evidence becomes available. Yet, at the same time, it may be that more precise prognoses of future climatic developments go unmade for political reasons; because, for example, Global Circulation Models with higher resolutions would allow the development of regional scenarios that might identify winners and losers due to climate change, thus posing additional difficulties for the international negotiations.

The case of ozone layer protection was different in that there was, before the consensus assessment reports, strictly speaking, no epistemic community. From the beginning, a few scientific advocates dared to combine their scientific work with practical judgement and with political recommendations or demands. Rowland was not afraid to demand first a ban on CFCs in spray-cans and then, after the discovery of the ozone hole, a general ban. Moreover, it was he who coined the metaphor of the ozone hole. His credibility and that of other advocates grew as time passed, particularly after the onset of dramatic events (the ozone hole). In the case of climate change this development was precluded by the deliberate creation of an epistemic community. Back in the 1980s, Stephen Schneider and James Hansen distinguished themselves as advocates of a policy of prevention. At public hearings, they did not hesitate to describe current extreme climatic events as expressions of anthropogenic climate change, for which they were much criticised (cf. Nance 1991). With the IPCC, this activity subsided. Climatologists thereby gained an exciting, relatively well-funded research field, but at a price: they could not move beyond the boundaries of the official consensus. This had two important implications. First, the consensus view and its propagation meant that the experts embarked on a course of presenting the 'mean risk'. By excluding worst case scenarios, the role of the public – which usually responds very well to alarming news – was minimised. At least this is the verdict if we believe research findings stressing risk aversion in the face of uncertainty – that the public 'irrationally' ranks the potential negative outcome higher than the potential positive outcome. The media, however, jumped eagerly upon extreme weather events, linking them to global warming and making them into harbingers of climate change. The IPCC distances itself from such scare stories which are said to be void of scientific evidence. Therefore, an important link that could have contributed to a self-reinforcing feedback between a risk-averse public, concerned scientists and affected governments of the globe has not been established.

Second, by the same token, sceptics and outsiders thereby seized the opportunity to deconstruct the available findings, which they did in public, in the mass media (see Gelbspan 1997). So in the end, all attempts at reaching a consensus view notwithstanding, debate and controversy could not be avoided. What is more, fierce enemies of regulation seem to dominate in the public debate (at least in the run-up to the Kyoto talks) where they are not attacked by equally adamant advocates of regulation but by a consensus view that expresses the least common denominator. The public thus is confronted with three sources and pieces of information: the IPCC which propagates a sort of 'mean risk' approach; the contrarian scientists who either attack global warming as bogus or as possible but beneficial; and the media which at times publishes worst case scenarios which, however, have little backing from the scientific community.

A popular explanation for the difference between the two cases cites the greater complexity of the problem of climate change, or how 'simple' it was to solve the problem of the ozone layer. In retrospect it may seem so, in accordance with a functionalist logic that declares solved problems to be easily solved problems. Upon closer examination, we can see that the case was anything but simple. For almost twenty years, producers of CFCs throughout the world resisted regulation, in part by means of the same arguments that are still heard in the case of climate: there were, they claimed, no cost-effective alternative technologies.[3] These were found when the producers were forced to forgo the use of CFCs. The anti-regulation position was still so strong in 1987 that six months before the signing of the Montreal Protocol, Lang, then chair of the international ozone negotiations, claimed that no more than 10 to 20 per cent CFC reduction was feasible in the next decade (*New York Times*, 28 February 1987).

There is some truth to the complexity thesis with regard to the structure of business in both fields. While Du Pont was the market leader, and its change of direction set off a chain reaction in the case of the ozone layer, this has not occurred in the climate case and it is doubtful if it can. Here, there is no dominant producer from whom all others take their cue. However, there are signs that the oil industry is giving up its obstructive role. In May 1997, during a lecture at Stanford University, John Browne of BP America announced that the company was in favour of gradual reductions in CO_2 emissions:

The time to consider the policy dimensions of climate change is not when the link between greenhouse gases and climate change is conclusively proven, but when the possibility cannot be discounted and is taken seriously by the society of which we are part. We in BP have reached that point.

BP's declaration could in any case be seen as a sign that oil producers no longer see their future exclusively in terms of oil and trigger a bandwagon effect.

But the important difference between the two cases is as follows. The ozone controversy was decided once a clear alarm signal had appeared and was used by a strong, publicly visible policy network in order to advocate strict controls. In the climate controversy, however, it remains unclear whether various extreme climatic events are at all connected with the long-term global climate changes. A few climatologists make this claim, but this does not constitute the consensus of the IPCC. In a way, the early institutionalisation of the epistemic community in the form of the IPCC suppressed any open controversy. In order to preserve a consensus (of which too much was expected politically), the scientific controversy was silenced. This gave outsiders the chance to make their name in the media, albeit being condemned as essentially unscientific by the mainstream epistemic community. If all conflicting opinions would have been openly aired, then large parts of the public could probably have been convinced of the serious implications of climate change and the need for a precautionary policy.

The construction of the IPCC as an international epistemic community committed to a scientific consensus and serving a political end has proven, on this view, to be somewhat counterproductive. The pressure to come to a consensus robbed the controversy of an essential dynamic. The symbolic interpretation of dramatic events, and the mobilisation by such means of third parties for the purpose of climate protection, have not been possible so far in the climate debate. This is a speculative lesson which follows from the above analysis. It rests on a counterfactual and therefore cannot be proven. If plausible, it would put into question the main lesson drawn by the architects of the IPCC.

The negotiations and control measures

I shall now dedicate some attention to the negotiation process and try to apply what I take to be essential lessons from the ozone case. To recap, normal bargaining and technical problem-solving are not sufficient for actors to solve conflicts since their cognitive orientation prevents them from doing so. Furthermore, the outcome of the international negotiation process is to a large degree determined by the balance of power between advocates and opponents of regulation within each country. The first have a common good orientation, the latter a narrow economic one. In the case that the common good orientation prevails, the opportunity arises to overcome narrow economic interests. Granting exceptions and transition periods for laggards and draggers would thus be the way to arrive at a compromise.

The international process of negotiation was directed by the Intergovernmental Negotiating Committee (INC), whose chair, Estrada, outlined the intended goal of a framework convention on climate in this way:

> It would not be profitable to draft a treaty that lays out highly specific policies that only a few countries could agree to. Rather, they drafted a general treaty that sets an overall framework within which all governments can work together. The treaty offers governments a well defined process for agreement, step by step, on specification.
>
> (cit. in Tolba 1998: 95)

The primary result of this treaty is a separation between framework agreement and protocol that allows nations to be brought into the negotiating process without obligating them to specific measures. First, a programme of support for research was agreed to. This means that the negotiating partners gain each other's trust; it also, however, makes it difficult for them to evade the findings of the scientific research. The hope is that, as in the ozone case, new dramatic findings will be made that can be used as a basis for tightening regulations.

Scientific consensus, however, does not automatically lead to corresponding regulations. In 1992 in Rio, a Framework Convention on Climate Change was passed, and in 1997 a protocol was worked out in Kyoto. In the protocol, the industrialised nations agreed that they would reduce their emissions of greenhouse gases by 5 per cent based on 1990 levels by 2008 to 2012. The countries with the highest emissions committed themselves to a reduction of 6–8 per cent. This is indeed a step forward; as yet, however, the promised goals fall far short of those presented by the scientific community (60–80 per cent).

Although no consensus has yet been reached on what the level of stabilisation should be, IPCC calculations make it clear that emissions must be reduced well beyond the goal agreed to in Kyoto to keep global CO_2 concentrations at a level two or three times the pre-industrial average. To be sure, no one knows whether this would be a 'safe' concentration. It is interesting to note that, in contrast to the ozone case, no one has proposed a return to the level of pre-industrial concentrations. In any case, the higher emissions expected in the coming decades demand correspondingly more drastic reductions in the future. Nonetheless, Kyoto can be seen as the point of departure towards ever more ambitious goals, much as in the case of the Montreal Protocol. Ambitious goals in terms of reduction send market signals that could lead to investments in R&D towards sustainable energy paths and ultimately to technological breakthroughs.

Two major conflicts have marked the climate negotiations: on the one hand the conflict between the USA and other industrialised nations, and on the other the conflict between industrialised and developing nations. In the USA, a fundamental reduction of fossil fuels seems to be impossible for reasons of domestic policy. As a result, the Americans prefer a solution that will provoke as little resistance as possible in their own country. Initially joint solutions on the international level were avoided in favour of propagating efforts on the national level.

Then the USA backed flexible mechanisms, such as 'Joint Implementation' (JI), 'negotiable emission quotas' (pollution certificates) and 'carbon sinks', without committing itself to specific reduction targets. The idea of Joint Implementation is based on the cost-effectiveness of climate protection measures beyond the borders of the developed industrial nations: the marginal costs of emitting one tonne less CO_2 are much lower in poorly developed economies than in economies that have already invested in such measures (going from coal to oil to gas, by increasing energy efficiency, by means of renewable energy resources). Developing nations, however, see this instrument as a means by which the wealthy nations can buy their way out of any obligation to do anything themselves. The industrialised nations are criticised for wanting to maintain their high standard of living even while they deny the developing nations the opportunity for economic development. Up to now JI has been carried out as a pilot project; the countries involved have made no commitments regarding goals for reduction. Even if JI does not lead to any drastic CO_2 reduction, learning effects can be expected, as well as the formation of institutional connections (solidarity) across national boundaries. This could form a basis for agreement on a comprehensive solution to the problem.

The second and third mechanism stress the global dimension of the problem. The argument is that it is a matter of reducing the total amount, not of set reduction targets for individual (or blocs of) countries and that market mechanisms are more effective because of the low transaction costs compared to state-imposed regulations. A system of negotiable quotas nonetheless requires binding reduction targets. Likewise, carbon sinks seem to make sense from a global viewpoint but appear to create 'loopholes'.

In the 1980s the European Union (EU) had already chosen to pursue a 'no regrets' strategy of increasing energy efficiency, and hoped to move the USA into making the same step (on the leading role of the EU in matters of climate policy, see Sbragia 1999). Otherwise, the marginal costs for Europe would be even higher than for the USA. For this reason, the EU supported binding reduction goals (targets and timetables) and rejected negotiable quotas. Such quotas would allow the USA, for example, to purchase the CO_2 quota of other countries (as in the case of Russia's 'hot air' after the collapse of the Russian economy).

It would be only a slight exaggeration to say that during the climate negotiations, the EU and the USA replayed the ozone negotiations with their previous roles exchanged. Here, too, the logic came into play that the 'weaker' party has to wait for concessions from the stronger, wanting in any case to keep the negotiations from breaking down. The USA managed essentially to force their position through, although they had to agree to binding reduction goals when it became clear that unilateral initiatives (or declarations of intent) had not led to any reduction. It remains to be seen whether – as in the ozone case – once a binding catalogue of measures has been agreed upon and ratified, a dynamic that leads to more stringent measures will develop.

The conflict between the industrialised and developing nations is fundamentally a conflict over the financing of the costs of reduction. Both sides agree that the industrialised nations bear the historical responsibility for the present situation and that they also possess the technical resources to help the developing nations reduce their emissions. They could not, however, come to agreement on how much compensation the developing nations ought to be paid in addition to the existing financial aid. Developing countries made it clear that they would allow the negotiations in Rio (1992) to founder if the principle of additionality was not enforced. The developed nations claimed in response that they are neither morally nor legally obligated to help the developing nations protect global environmental goods, thus countering the principle of additionality with the principle of conditionality. If assistance was to be paid, then it should be used only for the protection of *global* environmental goods; not, however, for economic development or national environmental projects. The developing nations saw in this the approach of a new form of attack on their sovereignty, an eco-imperialism. Here, too, the controversy continues, although it was not prominently debated at the last round of talks in the Hague.

Technical problem-solving

A purely technical solution to the problem was by no means feasible, because the parties involved know what effect a supposedly 'neutral' technical yardstick or standard would have on the interests (and interest groups) of their own countries. Nations that now already emit very little carbon dioxide would be hit harder by uniform reductions (or by means of an energy tax within the EU) than nations with high emissions, since the former had already taken action in the past. Countries with many inhabitants would profit from a regulation that calculates the emission indicator per capita, economically weak nations from a national aggregate indicator. Developing countries see themselves cheated of the possibility to catch up with the economic growth of the rich Northern hemisphere. Clearly, a complex of considerations of fairness is revealed here that cannot be expected to be solved with the aid of Rawls's veil of ignorance (Rawls 1971). Rawls's solution has become impossible precisely because we know *too much*.

One technical yardstick that could be established was the Global Warming Potential (GWP) which was constructed along the lines of the Ozone Depletion Potential and Chlorine Loading Potential.[4] As in the case of the ozone protocol, a basket of various substances was defined so that each country can reach the intended reductions by its own means. However, it will only be effective if countries realise how they will be affected by it, then bargain again about reduction quota, trying to obtain exceptional clauses, and so forth until a zone of agreement opens.

It is not surprising that Europe wanted to move the USA to tackle domestic CO_2 reduction. Together with Japan, it proposed to include only the three main gases (CO_2, CH_4 and N_2O) into the basket while the USA wanted to include SF_6, HFCs and PFCs. The USA pushed through their position to include six gases, which means that they need to do less in terms of real CO_2 reduction. Interestingly, some countries like Germany did not make as much use of HFCs as the USA in order to fulfil its Montreal Obligations. The USA thus has the option to switch to hydrocarbon technology (in refrigeration) in order to reduce its green-house gas reduction targets.

Comprehensive problem-solving

A comprehensive solution to the problem requires that hegemony be achieved by the group of actors orientated to the common good and ready to make conces-sions and exceptions to holdouts in order to bring them on board in due time. In the case of climate, this entails that OECD nations make a start and commit themselves to measures before demanding them of the former socialist countries and the developing nations. This is taken into account in the FCCC. However, there remain differences of opinion among the OECD countries regarding the instruments and goals of climate protection, so that the important question of exceptional rules for laggards could only be broached tentatively. At any rate, in Kyoto it was possible to reach an agreement on an average total reduction goal of 5 per cent, with individual countries setting different goals on a case-by-case basis. Hence many countries are permitted to emit up to 10 per cent *more* (see Table 8.1), a path already taken in the regulation of CFCs. This is an indicator that the logic of the comprehensive solution has begun to take effect.

Even the goal of a 5 per cent reduction will prove difficult, because emissions of greenhouse gases have increased considerably since 1990 and will probably con-tinue to increase unless radical changes in the worldwide production of energy and habits of consumption occur. Estimates are based on the assumption that the USA, in a 'business as usual' scenario, will by 2010 have increased its emissions relative to 1990 by 34 per cent (US Department of Energy, see World Resources Institute, www.igc.org/wri/wr-98-99/kyoto.htm).

The Kyoto Protocol thus represents only a first step towards the original goal of the FCCC, namely a stabilisation of the concentrations of greenhouse gases at a level that poses no danger for the climate system. Even if the Protocol is ratified

Table 8.1 Target greenhouse gas emission reductions by 2012, Kyoto Protocol. Per cent change from 1990 emissions

Australia	+8	Monaco	−8
Bulgaria	−8	New Zealand	0
Canada	−6	Norway	+1
Croatia	−5	Poland	−6
Estonia	−8	Romania	−8
European Union	−8	Russian Federation	0
Hungary	−6	Slovakia	−8
Iceland	+10	Slovenia	−8
Japan	−6	Switzerland	−8
Latvia	−8	Ukraine	0
Liechtenstein	−8	United States	−7
Lithuania	−8		

Source: United Nations (UN), Kyoto Protocol to the United Nations Framework Convention on Climate Change, Article 3, Annex B (UN, New York 1997).

and implemented (which is uncertain), its effect will be only to slow the increase in concentrations, not to halt it.

Climate change and beyond

Let me take up again the comparative dimension, the issue of how variations in risk debates and regulations vary across countries. In Chapter 1, I mentioned that within Europe a technocratic model of policy making prevailed that was based on a consensual political style. Expert committees meeting behind closed doors and giving advice to governments epitomise this practice. On the other hand, the USA is seen as taking a confrontational style with extreme positions clashing in public. This different institutional structure seems to favour a lower degree of risk aversion in Europe since an approach that aims at expert consensus tends to screen out public anxieties. The American institutional design, on the other hand, allows for worst case scenarios and thus for public involvement in an adversarial process. Since people tend to place a greater weight on worst case scenarios and thus behave risk averse, it would follow that the USA in general would adopt a more risk averse policy.

This is largely what we observed in the ozone controversy. However, the picture started to change in the mid-1990s when for reasons of domestic politics, the Republicans winning a majority in the Senate, thus putting pressure on the Democrat-led government, the Clinton administration became very wary of the challenge. Since CFC regulations amounted to economic burdens for private households (above all for retrofitting air-conditioning appliances) and greenhouse gas regulations would further increase them, the government tried to minimise these. As a consequence, the government asked manufacturers of CFCs to continue production as long as legally feasible in order to spread the period of transition (Smith 1998). Likewise, in the climate issue the government wanted to reduce the risk of falling into public disrepute by imposing a carbon tax. Europe, on the contrary, has been quicker in phasing out CFCs and taking its commitments seriously, a fact that could be explained by the lower transaction costs of consensual political systems during implementation. The reasons for Europe's stricter stance on climate policy are another matter.

However, if we look further afield, there may be anomalies that cannot readily be explained in this way. The cases of the growth hormone BST and genetically modified foods show that the USA has introduced these without much public attention, whereas the Europeans have started the bandwagon to revise a process which has led to the quiet and piecemeal introduction of such products and technologies. How can this be explained? First of all, there is clearly a learning process going on which takes different forms in different political systems. While industry in the USA has learned to tinker with the adversarial process and ultimately use it to its advantage, in Europe public interest groups have sprung up which engage in the representation of diffuse interests. This leads to the paradoxical outcome that – against theoretical expectations – Europe might end up with stricter standards in food safety than the USA.

This, however, is only a paradox if we place too much stress on institutional opportunity structures and neglect the power of policy networks. As events in the last years have shown, also in consensual and corporatist policies environmentalist and consumer groups can gain influence, partly by gaining direct access to the political system through parliamentary or government seats, partly by interacting with the media, and partly by receiving support from advocacy scientists. The resulting self-reinforcing loop may be so strong that institutionally ingrained policy traditions are challenged overnight. And it could well be that those institutions that had excluded citizens' concerns most decidedly are prone to suffering the most from the public backlash. It is the task of future research to delve deeper into these issues.

Appendix

List of interviewed experts (position at the time of interview)

Daniel Albritton	Director, Aeronomy Laboratory, National Oceanic and Atmospheric Administration, Boulder
James Anderson	Harvard University, Department of Chemistry, Cambridge
Richard Benedick	Ost-West-Wirtschaftsakademie, Berlin
Rumen D. Bojkov	Special Adviser for Ozone and Global Change to the Executive Heads of WMO and UNEP, Geneva
Holger Brackemann	Umweltbundesamt Berlin
Guy Brasseur	Director of Atmospheric Chemistry Division, National Center for Atmospheric Research, Boulder
H. Bräutigam	Solvay-Fluor, Hannover, Germany
Laurens Brinkhorst	European Parliament, Brussels
Ralph Cicerone	University of California, Irvine, Department of Geosciences
Paul Crutzen	Max-Planck-Institut für Chemie, Mainz
David Doniger	Council to the Assistant Administrator for Air and Radiation, US EPA, Washington, DC
Dieter Ehhalt	Kernforschungsanlage Jülich, Institut für Atmosphärische Chemie
Joeseph Farman	British Antarctic Survey, Cambridge, UK
Monika Ganseforth	MdB, Member of Enquete-Kommission Schutz der Erd-atmosphäre
Wolf-Dieter Garber	Umweltbundesamt Berlin
Hartmut Graßl	Director, Max-Planck-Institut für Meteorologie, Hamburg
Gerhard Hahn	Bundesministerium für Forschung und Technologie, Bonn
Neil Harris	European Ozone Research Coordinating Unit, Cambridge, UK
Jim R. Holton	University of Washington at Seattle, Department of Atmospheric Sciences
Herwig Hulpke	Bayer AG, Leverkusen
Heinrich-Wilhelm Kraus	Bundesministerium für Umwelt, Naturschutz und Reaktorsicherheit, Bonn
Karin Labitzke	FB Geowissenschaften, Freie Universität Berlin
Winfried Lang	Diplomatic representative of Austria at the UN, Geneva
James Lovelock	Coombe Mill, UK
Michael McElroy	Harvard University, Cambridge, MA.

Mack McFarland	Principal Consultant – Environmental Programs, E.I. DuPont De Nemours, Wilmington, Delaware
Jerry Mahlmann	Director, National Oceanic and Atmospheric Administration, Geophysical Fluid Dynamics Laboratory, Princeton University
Werner Maihofer	Former Secretary of the Interior, Germany
Peter Mencke-Glückert	Former head of ministry department, BMI
Alan Miller	Executive Director, University of Maryland, College Park, Center for Global Change
Mario Molina	MIT, Department of Earth, Atmospheric and Planetary Sciences and Department of Chemistry
Edda Müller	Secretary for Nature and the Environment of Schleswig-Holstein, Kiel
Michael Müller	MdB, Member of the Enquetekommission Schutz der Erdatmosphäre, Bonn
Franz Nader	Verband der chemischen Industrie, Frankfurt a.M.
Gerhard Pfleiderer	Hoechst AG, a.D.
Michael Prather	University of California, Irvine, Department of Geosciences
Sherwood Rowland	University of California, Irvine, Department of Chemistry
Rolf Sartorius	Umweltbundesamt Berlin
Ulrich Schmidt	Kernforschungsanlage Jülich, Institut für Atmosphärische Chemie
Friedhelm Schmidt-Bleek	Wuppertal-Institut für Klima, Umwelt, Energie GmbH
Mark R. Schoeberl	Goddard Space Flight Center, Greenbelt, Maryland
Steve Seidel	Director, Stratospheric Protection Division, Office of Atmospheric Programs, Environmental Protection Agency, Washington, DC
Susan Solomon	National Oceanic and Atmospheric Administration, Aeronomy Laboratory, Boulder, CO.
Richard Stolarski	Goddard Space Flight Center, Laboratory for Atmospheres, Greenbelt, Maryland
Clemens Stroetmann	Former Permanent Minister, BMU
Mostafa Tolba	President, International Center for Environment and Development, Geneva; Former Executive Director of UNEP
Ka-Kit Tung	University of Washington at Seattle, Department of Applied Mathematics
Tony Vogelsburg	E.I. DuPont De Nemours, Wilmington, Delaware
Robert Watson	Director of White House Office and Science and Technology Policy, Washington, DC
Steven Wofsy	Harvard University, Department of Earth and Planetary Physics, Cambridge, MA
Donald J. Wuebbles	University of Illinois at Urbana-Champaign, Department of Atmospheric Sciences
Reinhard Zellner	Universität GH Essen, FB Chemie

Notes

1 Social science and global environmental problems

1 For the notion of transnational relations, see Kaiser (1969) and Keohane and Nye (1971). It denotes regular interactions beyond national boundaries which include at least one non-state actor. In contrast, international relations are relations between states (Risse-Kappen 1995: 3).

2 Sachs *et al.* (1998: 208–9) make the point that the real test is yet to come. The authors see two critical points: the reluctant position of the USA regarding payments to the multilateral fund for the support of developing countries to make the transition to CFC alternatives, and the willingness of the developing countries to take serious measures after the 10-year grace period has come to an end. According to UNEP, the USA has fulfilled all its duties in this respect (Mani Subramanian, Multilateral Fund Secretariat of UNEP, personal communication, 14 July 1998).

3 There are indications that their implementation can be considered successful, too; see Montzka *et al.* (1996) and Parson and Greene (1995). According to the United Nations Environmental Program (UNEP), world production of chlorofluorocarbons was virtually cut in half during the period from 1986 to 1992 (UNEP 1994: 32).

4 A large *k*-group would reduce the chances for successful cooperation even where the logic of a prisoner's dilemma applies (Hardin 1982: 153, 193). However, in absolute numbers there were few countries with domestic CFC production (cf. Downie 1995: 334). For a discussion of the influence of the number of actors on the outcome of negotiations see Keohane and Ostrom (1994).

5 See Krasner's definition of an international regime as 'implicit or explicit principles, norms, rules and decision-making procedures around which actors' expectations converge in a given area of international relations' (Krasner 1983: 2).

6 Compared to prisoner's dilemmas, deadlocks are much more resistant to solution since each side prefers to defect instead of cooperating. The logic of deadlocks has rarely been investigated in the literature on international relations. See Axelrod and Keohane (1986), Downs *et al.* (1986).

7 In their analysis of arms races, Downs *et al.* (1986) have found three factors that can help in dissolving deadlocks: the intervention of third parties, issue linkage, and changes in economic context. The role of third parties is an important one for this study.

8 I do not take up the literature on decision theory (cf. Luce and Raiffa 1957: 284ff). See Elster (1989: 84–5) for a critical view.

9 There is only one world history. Historians who try to make scientific statements sometimes use counter-factual thought experiments and theories of possible worlds (Elster 1978). Atmospheric scientists seem to solve the methodological problem in a similar way. This poses interesting questions about the relation between natural and social sciences which can only be mentioned in passing.

10 See also the following definition of discourse coalition: 'Discourses are open systems of communication which "go back and forth", get interrupted and are rekindled at other places. There are no exclusive membership roles' (Evers and Nowotny 1989: 361).

11 The notion of ideology has mainly political connotations. To avoid misunderstanding, I shall make clear that by ideological orientation I mean cognitive orientation in the sense of a mental map or a cognitive frame.

12 The knowledge produced in public administration (regulatory science, cf. Irwin *et al.* 1998; Jasanoff 1990) seems to command a lesser prestige than the knowledge produced by large autonomous research institutions.

13 Hall (1989: 362) observed that 'It is ideas, in the form of economic theories and the policies developed from them that enable national leaders to chart a course through turbulent economic times, and ideas about what is efficient, expedient, and just that motivate the movement from one line of policy to another'. Keynes has also stressed the power of ideas: '[T]he ideas of economists and political philosophers, both when they are right and when they are wrong, are more powerful than is commonly understood. Indeed the world is ruled by little else. Practical men, who believe themselves quite exempt from any intellectual influences, are usually the slaves of some defunct economist. Madmen in authority, who hear the voices in the air, are distilling their frenzy from some academic scribbler of a few years back. I am sure that the power of vested interests is vastly exaggerated compared with the gradual encroachment of ideas' (Keynes 1936: 383).

14 Norms are real because they have an independent motivational power; they are autonomous since they cannot be reduced to rules of optimisation: 'There is no single end – genetic, individual or collective – that all norms serve and that explains why there are norms. Nor, for any given norm, is there always any end that it serves and that explains why it exists' (Elster 1989: 125).

15 An interesting aspect regards the value of the past. Normally, it is assumed that rational actors forget about sunk costs and look forward to the future. In path dependent processes with antagonistic constellations, the actors do not simply write off their investments in the past – too much is at stake for them (Elster 1989: 98–9; Wolf 1970). However, once one adopts this logic – and this is often the case in public debates where one is being watched by third parties (Hirschman 1982: 79) – as time goes by it becomes more and more difficult to escape from it. The investment into a possibly lost cause increases in the hope of reaping the benefits eventually, but it may be too late to note the disastrous consequences.

16 Here I follow the terminology of Young who distinguishes between distributive and integrative bargaining. In the former, negotiators know the shape of a welfare frontier and will therefore 'turn to calculations regarding strategic behaviors or committal tactics that may help them achieve their distributive goals' (Young 1994: 100). In the latter, negotiators 'do not start with a common understanding of the contract curve or the locus of the negotiation set' and therefore have a strong incentive 'to engage in exploratory interactions to identify opportunities for devising mutually beneficial deals' (Young 1994: 101).

17 Murphy in fact seems to subscribe to a rather naive view about our sensory capacities to perceive environmental problems (see Grundmann 1999).

18 In his study on the German chemical law, Volker Schneider observed that the thematising by science 'has not been sufficient to put the topic onto the government's agenda. ... An important additional factor is the public's non-acceptance of a situation or a problem'. It may be the case that this protest is launched from within the government or from international organisations (Schneider 1988: 188).

19 Scientists obtain data that have been produced with the help of instruments and machines from laboratories (Shapin 1984). This lends them an objectivity that seems far above the judgement of other citizens, politicians, journalists or managers (Porter 1995).

20 This sociological description of knowledge production as a social process is straight-foward and also applies to fields like mathematics in which one speaks about logical proofs and refutations that seem to exclude the social dimension. However, before proof becomes accepted as proof, it has to be accepted by the community of mathematicians (MacKenzie 1993).

21 'Precedent seems to exercise an influence that greatly exceeds its logical importance or legal force. A strike settlement or an international debt settlement often sets a "pattern" that is followed almost by default in subsequent negotiations.' This is especially the case if the precedent is highly visible (Schelling 1960: 67f.).

22 In the 20-year period, two scientists in my sample moved from the USA to Europe, one went in the opposite direction.

23 I encoded the interview sources in the following way. The first two letters indicate the geographic–institutional location (US, UK, GE, UN, EC), the last two letters, the field of activity (AS denotes atmospheric sciences, AD administration, PO politics, IN industry, DI diplomacy and EN environmental group) and the numbers refer to the whole sample of interviewees. Thus USAS refers to an atmospheric scientist from the USA, UNDI to an UN diplomat, GEPO to a German politician, and so on.

2 Ozone science

1 The relative proportions (in volumes) are as follows: nitrogen 78.11 per cent; oxygen 20.95 per cent; argon 0.934 per cent; carbon dioxide 0.035 per cent. The concentration of water vapour fluctuates. In comparison: ozone *c.* 0.000001 per cent to 0.000003; CFCs 0.0000003 per cent; natural chlorine concentration 0.00000006 per cent (Graedel and Crutzen 1995).

2 This is a simplification. Polar night is latitude dependent. Only at the South Pole does the sun not appear until the autumnal equinox. The losses of ozone within the Antarctic ozone hole begin around late August – a month before the equinox – because sunlight has begun to penetrate these latitudes. At 70 degrees South, there are 7 hours and 20 minutes of daylight at the surface on 15 August; at 80 degrees, zero on 15 August, but almost 11 hours of surface daylight on 15 September, and of course 24 hours on 15 October. I am grateful to Sherry Rowland for pointing this out to me.

3 The reservoir chlorine nitrate ($ClONO_2$) forms out of two gases which are ozone depleting substances, ClO and NO_2.

4 Without this catalytic amplification, the process would not proceed as fast as it does. This element of the explanation, the so-called 'ClO-dimer' (Molina and Molina 1986) is today seen as the decisive theoretical contribution in explaining the Antarctic phenomenon.

5 No matter which key words are chosen as indicators, we only get an approximation to the real development of the field, since key words in the *SCI* are picked partly by the scientists themselves and partly by the journal editors (personal communication Eugene Garfield). This leads to two problems: first, both may be different, and second, there may be self-reinforcing loops. This is highly probable in cases where the field of research has a high communicative density. Researchers publish their findings under the same key words as their colleagues. The volatility of key words thereby prevents an adequate description of the quantitative development of the field since new key words may emerge at any time.

6 Cf. Stinchcombe (1984) for a similar view on sociology. In his view, theoreticians erect a dam to protect themselves against the flood of 'dirty data'.

7 A scientist who belongs to the core group of the scientific community put it this way: 'There were three papers, where theory people tried to explain the ozone hole. They all take credit for it, but they were all wrong. The mechanisms were all wrong. S. had HCl which can't do it, M. had BrO + ClO which is 20 percent, but he can't make the ozone

hole, and C. had some HO_2NOH system. That model turned out to be irreproducible. They all had heterogeneous chemistry, in that sense they were all right. But they were all wrong in the sense that no one can get the ozone hole. If you go back and look at the papers, they are all embarrassing, they all have kind of black magic; everyone tried to get the ozone depletion right, but nobody could do it. The first person who could do it was Mario Molina, who did lab work which showed that the ClO dimer could form, combine, photodissociate, destroy ozone at a rate that was kind of what we have seen. If you put that into the models, they started to give the right answers. Farman observed it, everyone said it's heterogeneous chemistry, they all jumped on the bandwagon, but even then they couldn't come up with a mechanism' (USAS 15).

8 'What would drive a group of people to do as much as scientists do in their jobs, to work 80-hour weeks and . . . you know, I mean there are a lot of people in this field who work insane hours. Competitiveness must be in it, otherwise we would have the good sense to enjoy life more' (USAS 45).

9 Two dynamicists whom I interviewed said that the workshop in Snowmass brought the chemists' victory.

10 Chubin and Hacket's data show (1990: 66) that a large number of scientists (60 per cent) believe that the peer review system blocks innovative ('unorthodox or high risk') approaches.

11 The concentration is measured by the proportion of molecules within an air sample. The units are ppm (parts per million, 10^{-6}), ppb (parts per billion, 10^{-9}) and ppt (parts per trillion, 10^{-12}). Since these are proportions of volumes, they are also abbreviated as ppmv, ppbv and pptv.

12 In 1978, the US Bureau of Standards carried out a survey of 16 principal laboratories in the world that were measuring CFCs (Hughes *et al.* 1978). They found that they differed by more than 50 and 100 per cent for CFC 11 and 12 respectively. This high level of disagreement underlines the scientific uncertainty about CFC concentrations in the first decade. Given this degree of discrepancy, it was almost impossible to calculate residence time of CFCs accurately.

13 The following numbers are based on an analysis of the bibliographies of the WMO reports of 1985 and 1994.

14 If one compares the rate of cooperation in the British natural sciences (Hicks and Katz 1996: 390) with that in the field of international ozone research, the latter was slightly below the British average in 1985 and far above the British average in 1994.

15 The recognition of NASA as the leading coordinating scientific institution is due to its high prestige. It was hoped that its uncontested 'seal of approval' would make foreign countries accept the research results. Put in neo-institutionalist terms: 'Absent coercion, the other parties will be willing to delegate such discretionary authority only if they believe that it will be used fairly and effectively. An important source of this belief is reputation. The party to whom authority is delegated should be the one with the most to lose from a loss of reputation' (Majone 1996b).

16 Crane and Price take up a notion that Robert Boyle had coined in the seventeenth century which refers to 'past and present informal collectivities of closely interacting scientists limited to a size that can be handled by inter-personal relationships' (Merton 1995: 407).

17 The importance of these scientists was measured by canvassing specialists in the field. A first sample of interviewees was drawn on the basis of literature research and then extended by snowball sampling. A good indicator of their central importance is the fact that all of the scientists interviewed served at least once as an author or reviewer of a chapter in the international WMO reports. In addition, about half of them were responsible for individual chapters. The sample also includes all the authors of the most important publications during the period of the survey, as well as the three winners of the Nobel Prize for Chemistry in 1995. Although the results are not representative for the field of ozone research, they are meaningful because a number of the

scientists interviewed took a central advisory role in the policy process and in the public debate.

18 It is relatively easy to classify vocal scientists since they left their mark in newspapers, public hearings, and so on. Quiet scientists are more difficult to locate. They do not speak out in public but engage in policy advisory activities. Here I relied on my interview material.

19 I am grateful to Susanne Hilbring for her help with the database research.

20 This statement refers mainly to the US since in Germany the controversy gets going only in the 1980s. At the end of the 1970s, one of the early advocates moves from the USA to Germany and takes on the role of a vocal advocate, albeit not before 1988.

21 In the 1970s, one of my interviewees was a referee for a report of the National Academy of Sciences. He remained undecided about the question of regulation.

22 Of course, this is the current opinion of the chemists. During their professional training, they learn in physical chemistry about the diffusive de-mixing of the atmosphere, the barometric layering of the molecules. According to this, 'the stuff cannot go up, but the point is that the atmosphere is mixed turbulently and diffusive de-mixing takes place only above 110 km' (GEAS 26; cf. Rowland 1993; WMO 1994: xxv–xxxiv).

23 In the US Congress there was a fierce controversy about this (cf. Brown 1996). The Tindall report analysed US television news and found a decrease in coverage about environmental problems from 1989 to 1993 by about 60 per cent. Kevin Carmody (1995) analysed the print media and concluded that since Earth Day 1990 there is less and less interest in environmental topics, even in serious papers like the *Economist*, the *New Yorker*, the *Wall Street Journal, Los Angeles Times* and the *New York Times*. Typical headlines read: 'Are we scaring ourselves to death?', 'Living scared: why do the media make life seem so risky?', 'Environmentalists are on the run: business leaders, local officials, and angry citizens are demanding an end to rules based on silly science and bad economics'.

24 'There is something about the rewards for those people. It does not lie with the science but with being sceptics of the science. . . . It is [their] lifestyle. And there is a political agenda underneath of it. [In their view] industry needs being promoted and humans can't really affect the environment, you know. Dixie Lee Ray told me in person that she did not believe that there was any way that humans can affect the environment' (USAS 47).

25 On page 271 we read: 'Today many leaders of Germany's Green Party are former members or admirers of Hitler's SS.'

3 Ozone controversies

1 In contrast, it seems as if the social sciences and humanities like to invoke the authority of dead spirits. There is a high proportion of publications that deal with the patrimony of the classics. Price (1970) uses this as an indicator to distinguish between hard, soft and non-science. In the natural sciences the invocation of the dead spirits is not part of the core research but part of the historiography of the discipline.

2 For an early critique, see Barnes and Dolby (1970) and Mulkay (1969); for a defence of the position, see Ben-David (1991b) and Zuckerman (1988).

3 Some Mertonians have conceded that Merton's norms are discarded during scientific controversies. Thus Ben-David comments on some case studies on scientific controversies: 'The scientists at this stage act like litigants concerned more with putting together a convincing case than with ultimate truth. They are not, and are not expected to be, dispassionate' (Ben-David 1991b: 480). Mukerji states that this is the general condition in science (1989: 169): 'Passion and sovereignty more than detachment characterise the way scientists relate to the natural world (or their traces of it) in the daily life of laboratories.' See also Mitroff (1974).

4 This finding is different from Fuchs (1993a) who would expect severe sanctions of the scientific community.

5 One can construct this case as a purely inner-scientific dilemma. Imagine a scientist who makes a radical new finding. After some time she realises that her discovery was based on wrong assumptions which so far have escaped the scrutiny of her competitors. Should she publish the new evidence?

6 Carl Djerassi, the inventor of the anti-baby-pill, has written three novels about related aspects, drawing upon his intimate knowledge of scientific practices.

7 The reasons why both periods are not represented within one graph is that the database of both diagrams is different. In the first we have three advocates and two sceptics, in the second, eight advocates and four sceptics. The early sceptics are no longer active after 1986. Their place has been taken by other scientists who are sceptical about the relation between CFC and ozone hole. The early advocates are joined by other scientists.

8 Here I used only data of scientists who were actively involved in the controversy. The inclusion of bystanders would have led to different results.

9 United States Senate, Congress Hearings 18, 19 and 23 September 1975 (in the following cited as Congress Hearings 1975), 939, statement Rowland.

10 Congress Hearings 1975, 570.

11 Congress Hearings 1975, 1023, statement Anderson.

12 Congress Hearings 1975, 1048, statement McElroy.

13 An atmospheric scientist characterised the professional status of the NRDC as follows: 'They . . . had intelligent people, not necessarily particularly informed about the chemistry of the CFC controversy but willing to explore it. . . . They wanted to check out for themselves whether or not they felt that this was plausible. They probably required a lower level of plausibility than the NAS two years later' (USAS 16).

14 'As regards Greenpeace, and I have clearly some sympathy with environmental pressure groups, I think there is also the danger that they are shooting themselves in the foot. When you say something too strong which is easy to be proven to be too strong and you are therefore too weak in your argument, that is stupid' (UKAS 43).

15 Cf. the statement of a bystander: 'I had had to really steer clear of it. I had this idealistic view, I wanted to be seen as a seeker of the truth. I didn't let my prejudice overcome me' (USAS 45).

16 'The catalyst for the OTP was Don Heath testifying in front of Congress and various other places that the satellite data showed a very rapid decline in ozone globally. The basic feeling was that Don Heath hadn't done his homework and he was wrong. The result of the OTP, simply said, was: Don was wrong, but he was right. He was wrong because there was a calibration drift in his data, but indeed ozone was going down globally, by a much smaller amount than he was claiming' (USAS 45).

17 In 1986 there were reports in the scientific literature about problems with the SBUV (Solar Backscatter Ultraviolet Spectrometer) and TOMS instruments (Fleig *et al.* 1986).

18 Cf. the outrage of an American scientist: 'The satellites have very, very bad calibration. The Dobsons need better calibration but this refers back to the poor state of long-term measurements. But NASA has been rescued by the Dobsons. They are imperfect, but in order of magnitudes better than the satellites. The diffuser plate from the SBUV and TOMS instruments has a vicious degradation. . . . I am upset by that' (USAS 11).

19 Scientists are aware of the problem of commissioned research: 'I can't help it, his research has been funded by industry. One gets to know each other, is friendly and this may influence the results and the interpretation of the results. I think none of us is immune against this influence' (GEAS 25).

20 The other members of the OTP were Angell, Attmannspacher, Komhyr, McPeters and Stolarski. It was chaired by Rowland (WMO 1988: 179).

21 Readers familiar with Popper's terminology may be puzzled; an atmospheric scientist expressed the view of his colleagues in the following way: 'Verify is actually what we are

doing. We are testing things with datums . . . and a datum is a number that you believe in – if it's for a model or a measurement' (USAS 15).

22 The groups originated from NOAA's aeronomy laboratory in Boulder, from the University of Wyoming (Laramie), from the Jet Propulsion Laboratory in Pasadena and from the State University New York, Stony Brook.

23 'Jim Anderson was able to do that. He had a very robust instrument which was challenged by nobody' (USAS 27). 'The signal to noise ratio is very clear in Anderson's data. He can measure chlorine oxide at 10,000 times less than what he was seeing' (USAS 16). Galison (1987: 183) describes a similar credibility mechanism in the field of particle physics.

24 A classical example refers to the controversy between Hobbes and Boyle (Shapin and Schaffer 1985). Thomas Gieryn draws special attention to the boundary work employed by both and thus summarises Boyle's strategy: 'Success would be likely, if Boyle could move everyone – rivals, bystanders, audiences – onto his playing field. With borders that he drew and labelled. He did that in a crafty move, in effect arguing that only those who were in the experimenter's community . . . could challenge the claimed facts. But, Catch-22, the price of admission to the lab (and to the Royal Society, as Hobbes found out) was a commitment to Boyle's program' (Gieryn 1995: 428).

25 Natural scientists publish mainly articles in peer-reviewed journals. Since peer review plays a lesser role in book publications, they are more loosely coupled to the core of research. Scientists who mainly publish books are not taken seriously.

26 Haas is wrong in claiming that Farman had got his results from a recalculation of satellite data: 'In mid-1985 . . . Farman published an article describing a rapid and unexpected thinning of the ozone layer over Antarctica during the Antarctic springtime, based on a recalculation of existing satellite data' (Haas 1993: 157). Farman's measurements were completely independent of NASA. He used no satellite data at all, only the ground-based Dobson and the balloon-based ozonesonde and temperature measurements – all BAS data.

27 A similar experience was faced by a researcher of a major American university after he had been advocating CFC regulations: 'I think from a personal point of view, the seriousness of this issue has not been properly addressed by industry. And I said so . . . and that created rather an explosive situation between Du Pont and myself' (USAS 8).

28 'The trouble is there are very few people in the world who are able to handle a Dobson. Even Bojkov himself would not have understood what we have done to our data set. WMO had failed dramatically, the International Ozone Commission did a terrible job in keeping the world network' (UKAS 44).

29 This is confirmed by a modeller: 'The Dobson instrument is an old kind of instrument, it's very hard to operate and the operator has to be highly trained. And in the 1970s people were not that interested in the Dobson network until ozone loss came along. So a lot of these stations were not very well calibrated, and the operators were very low trained, so the data looked real rough' (USAS 17).

4 Country comparison

1 'When Rowland called Johnston in November or December of 1973 to tell him what they had discovered, Johnston told him that we should all get together. Rich Stolarski had met Hal Johnston at the meeting in Kyoto, Japan so he was able to refer our work to Molina and Rowland. But we didn't know each other, so we had to exchange letters very formally. After the publication of the Molina–Rowland hypothesis, Molina, Rowland, and we met at a scientific meeting in California. That was the beginning, from my point of view' (Personal communication from Cicerone).

2 'Industry group launches defense of fluorocarbons' (*Los Angeles Times*, 28 July 1975). Another scientist on Scorer: 'They brought him over with a public relations outfit, they

ran him around the country, in talk shows; as it happened, I got involved in some of those. My views may have been conservative at this stage but I certainly did not take the conservative position when I appeared in the William Buckley "Firing Line"show with Richard Scorer because it was clear to me that he was a phony and that this was not a serious discussion' (USAS 30).

3　When being asked how accurate his measurements were, Lovelock gave an estimate of 20 per cent. As he told me, one American scientist especially claimed to be able to measure CFCs to 1 per cent accuracy. It subsequently emerged that he confused precision with accuracy.

4　The labs that participated in the test were: R. Cicerone, University of Michigan; R.K. Stevens and A.I. Coleman, EPA; H.H. Gill, Dow Chemical Co.; R.A. Gorski, Du Pont; B.J. Tyson, NASA Ames; P. Fraser, CSIRO, Australia; S.A. Penkett, Harwell, UK; R.J. Lagomarsino, US Energy R&D Administration, NY; P.D. Goldon and T. Thompson, NOAA, Boulder; L. Elias, NRC, Canada; H.W. Singh, Stanford Research Institute, CA; H.I. Schiff, York University, Ontario, Canada; J.E. Lovelock, Bowerchalke; R.A. Rasmussen, Washington State University, Pullman, USA.

5　Letter to Rowland and Raymond McCarthy (Du Pont), 1 April 1975.

6　US House of Representatives, Committee on Interstate and Foreign Commerce, Hearings, 11–12 December 1974, *Fluorocarbons: Impact on Health and Environment* (Washington, DC: US Government Printing Office, 1974).

7　In September 1975, February 1976 and December 1976 hearings took place before the Subcommittee on the Upper Atmosphere led by Senator Bumpers (Roan 1989: 44–7). Bumpers himself made no secret of the fact that he favoured a rapid restriction on CFC propellants. An immediate ban was again supported by Rowland and Cicerone, this time with the backing of James Anderson. NASA head James Fletcher, on the other hand, was willing to accept further delays in order to obtain additional information.

8　To be sure, Molina and Rowland held to their original estimate of 7 to 13 per cent. They had come to the conviction that chlorine nitrate ultimately has no major influence on the ozone budget. Seven per cent was still considered by experts to be a serious threat; only at a value of 3 to 4 per cent ozone depletion was the problem no longer held to be serious. In the *Chicago Tribune* they were quoted as saying: '[chlorine nitrate] has never been proved outside the laboratory and has not yet been found in the stratosphere.' Their claim that chlorine nitrate was 'not particularly important' may well have amused the NAS committee, which at the time was desperately attempting to take account of the role of chlorine nitrate in its final report (Dotto and Schiff 1978: 258–9).

9　Several lead headlines used this onomatopoeia: 'Not with a Bang, but a Pffft', read the *New York Times*, 21 December 1975.

10　Sullivan's first article appeared on 27 September 1974. On the *New York Times* as opinion leader, see Elfenbein (1996).

11　The magazine *The New Yorker*, highly regarded for its extraordinarily carefully researched articles, must also be mentioned here. *The New Yorker* is one of the few periodicals that hires 'checkers' to confirm the research of its authors. It is here that Paul Brodeur published two frequently cited articles on the subject, 11 years apart (Brodeur 1975, 1986).

12　Medical and military applications, above all, were exempt. This was the first phase of a planned two-phase regulation. The second would involve 'essential' products. See Wirth *et al.* (1982).

13　The EPA and FDA passed ordinances mandating product information, and the FDA also demanded a warning label ('Caution – contains CFCs that can be hazardous to health or the environment due to ozone-destroying effect').

14　EPA (1987) Protection of stratospheric ozone, 52 Federal Register 47491.

15　'If you deplete ozone at 40 kilometres, it creates more ozone below. A lot of models cancelled that effect. So you have been really kicking the atmosphere, even in the 1981 models 2 per cent gain down here would cancel 10 per cent loss up there, because of more density, that's the way 1-D models worked for a while and that's why the net effect tended to be zero, or crossed zero. You were playing havoc with the atmosphere. Chemistry was changing all over the place, but you got this cancellation' (USAS 15).

16 In 1972, Lovelock had speculated that CFCs might be a greenhouse gas and reported the findings at a scientific meeting in Andover, MA. Du Pont, which paid him, did not try to suppress this finding (Personal communication from Lovelock).

17 'This seemed a sensible thing to do, had a large share on the market, it was a growing amount, it was direct emission to the environment and it was really a luxury type of item, there were competing technologies already there. They did not have to be developed, ranging from hand-pumps to different propellants' (USAS 13).

18 During the first hearings before the Bumpers Committee, they were also supported in their findings by reputable and influential scientists from other fields, as, for example, by Harvey Brooks: 'We have to prepare, if not to begin, limitation of the use of Freon for nonessential activities [such as cosmetics] or activities where there are relatively good substitutes' (US Congress Hearings 1975: 782).

19 The account in this section is indebted to Küppers *et al.* (1978), Müller (1986) and Timm (1989).

20 The German term *Minister* is translated as Secretary while the *Staatssekretär* is rendered as Permanent Minister.

21 'Based on the example of the USA, the German federal government also wanted, with its environmental programme of 29 September 1971 . . . to place environmental protection on a completely new basis . . . Both the American National Environmental Policy law . . . of 1969 and the establishment of a central American environmental authority (the EPA) were the blueprint for the strategic environmental considerations of the German government (Mencke-Glückert 1990: 243).

22 H.-J. Luhmann (1991) describes how the division into environmental media hindered the discovery and explanation of forest dieback (*Waldsterben*).

23 The literature sees this partly in a more positive light, for example, Müller (1986) and Timm (1989).

24 As sources for this section, archival material from the UBA, generously placed at my disposal, was utilised.

25 'The Du Pont upper management probably did not (and would not) trust the calculations of anyone else's model on a critical point such as this. So they put their own group together to make an "engineering model". . . . The first credible 2-D models were put together by Guy Brasseur and separately by Paul Crutzen in 1975. The Du Pont model came along about two years later, and obtained essentially the same results as the earlier models – confirming to the Du Pont management that the other results were believable. However, when Du Pont started sounding as though they believed that the models would be useful, it was a forward step for the modelling community because it removed one of the possible points of contention' (USAS 16).

26 One exception to this statement may be an article by two Hoechst employees in the specialist journal *Berichte der Bunsengesellschaft, Physikalische Chemie* ((82) 1978: 1147–50), in which the existence of tropospheric sinks is assumed.

27 See also Müller (1986: 80). On the basis of this finding, highly placed members of the Ministry of the Interior aided the establishment of environmental policy pressure groups (GEAD 12).

28 'There was a representative from the EPA (Barbara Blum) there who made an emotional speech, Permanent Secretary Hartkopf didn't like that at all. It sounded like a field sermon to him. That was clearly seen as inappropriate' (GEAD 7).

5 The road to Montreal

1 It is worth noting that all the nations involved had sent high-ranking representatives (permanent secretary or higher) to this meeting – with the exception of the Federal Republic of Germany.

2 'During the early 1980s the government of the United States under President Ronald Reagan undertook a reversal of national policy that can fairly be described as extraordinary. Appointees of the president sought to disengage the United States from international environmental policy commitments. Attempts were made to undo official involvement in programs and agreements in which the United States had often been an initiator', remarks Caldwell (1984: 319).

3 Reagan's successor Bush again revised these cuts; during his term of office the EPA budget was raised by 50 per cent and the number of staff by 22 per cent (Vig and Kraft 1994: 19).

4 'For about two years I was literally the only environmentalist who followed the ozone issue', the former NRDC representative told me. He came to an understanding with those responsible at the EPA in order to maintain the pressure on industry. 'There is a brief funny story of how alone I was. There was one hearing where industry was given a panel to present five witnesses. But I was the only one for the environmental case, but it would have been embarrassing to admit it. So I wanted to have three, phoned my friends and said: I write you a testimony, all you have to do is to come and read it. And they came. That was the desperate period.'

5 In the so-called Advance Notice of Proposed Rulemaking, a feature of the American policy process, a government agency announces goals for regulation, to be followed by a period of public discussion.

6 The cited NRDC activist recalls: 'He was considered quite honest. After he took over, we filed a petition in May 1983 in which we said: Look, you made this finding in October of 1980 that a continued build-up of chlorine in the atmosphere represented a risk. Under the Clean Air Act you have to either change your finding, (and we were confident that the science was going against that), or you have to take action.'

7 'I recognize that unilateral action has both advantages and disadvantages. It would provide an early incentive for our domestic industry to begin work on producing alternative chemicals. . . . By passing legislation in the United States, we would get the jump on other nations which I am convinced are going to have to take action. It is my understanding that such chemical substitutes are possible, but would take approximately five years to develop and would cost consumers a few pennies more. This seems like a small price to protect the ozone layer' (*Congressional Record, Proceedings of the 99th Congress, Second Session*, 8 October 1986: Senator Chafee).

8 At that time it was made public that the Reagan administration had exchanged weapons for hostages with Iran and secretly financed the Nicaraguan Contras with the profits ('Irangate' and 'Contragate'). Shultz, who had opposed these actions, enjoyed high popularity and was able to exercise particular influence in the cabinet.

9 'I've heard Watson say that that wasn't quite what Hodel said. He was misquoted, but when asked confirmed it [laughs]. It had gotten so far along that they decided it was too embarrassing to back out. The people in the President's cabinet did not find out that the decision was being made until it was too late' (USAS 16).

10 It should be recalled that at the time, the idea of banning an entire class of industrially produced chemicals by means of international measures was completely outlandish. I am grateful to Konrad von Moltke for this suggestion.

11 In 1989 a spokesman for Hoechst estimated the following degrees of potential substitutability: for coolants 80 per cent, for insulation 25 per cent, for aerosols 5 per cent, for cleaning agents 30 per cent and for soft foam 0 per cent (*Europa Chemie*, 13/89: 206). Du Pont, in the same year, estimated the following figures: Substitution by HCFC and HFC respectively 30 and 9 per cent, reuse 25 per cent and 'outside producers' 32

per cent. Four years later the assessments had changed in favour of substitutes: Du Pont still saw only 11 and 15 per cent each for HCFC and HFC, 29 per cent for reuse, but 49 per cent 'outside producers'. In the meantime it had turned out that over-capacities had been created and therefore a decline in the price of HFC had occurred (Brack 1996: 31).

12 'F 134a is considered as an alternative to F 12 for refrigeration and air-conditioning applications . . . and appears to be technically the most promising substitute among the fluorocarbons. It also appears to be a feasible aerosol propellant' (Umweltbundesamt 1989: 40). In 1989, the German Federal Environmental Agency reported that F 134a would not be available in commercial quantities prior to 1992.

13 Information according to *Europa Chemie* and *Chemische Industrie* (1988–90).

14 'F 141b has hitherto been produced only in small quantities, although a technical manufacturing process is known. . . . [It] is being tested as an alternative to F 11 for use in manufacturing polyurethane plastic foam . . . [and] for use in refrigeration technology' (Umweltbundesamt 1989: 38).

15 'DuPont was ahead of Allied in 1973, and were still ahead 15 years later, but the lead was about the same throughout' (personal communication with Rowland).

16 'I also do not have the impression that Du Pont had such a big lead in alternatives at the time. Back in the period 1977 to 1979, each of the major companies (Du Pont, Allied, ICI, perhaps Hoechst) obtained patents on particular methods for making some of the likely substitutes. . . . [The Du Pont decision to phase out] doesn't sound to me like a decision made because they thought that they had competitive advantage – although they may also have felt that they were in a good position' (USAS 16).

17 Even greater were the losses for the smaller producers who had no possibility of switching over to other products (F 22, 113) or moving to other countries.

18 A spokesperson for the German chemical industry sees in this one of the reasons for the 'failure' of the Töpfer environmental policy: 'Yes, that's certainly one reason for the failure [*sic*] of the Töpfer environmental policy, the fact that he separated that from industry' (GEIN 48).

19 The EC claimed that before Montreal it had been more willing to compromise than both of the other countries. This assessment is very likely the wishful thinking that, in a way, had led to the EC's being taken by surprise in Montreal when the USA came to an agreement with Japan and the Soviet Union, ruining the EC strategy of acting as mediator between extreme positions (USAD 23).

20 As mentioned earlier, the Austrian chief negotiator Lang thought it possible only to achieve a stabilisation at the level of 1986 and a reduction of 20 per cent three years after passing the treaty. Benedick, in contrast, considered such a solution totally 'unacceptable and ridiculous'. He stated that an 85 per cent reduction was needed just to prevent the further increase of current atmospheric concentrations. The German delegate signalled that the Federal Republic of Germany (like Denmark and the Netherlands) would support a reduction of 50 per cent.

21 Participants at the workshop included, among others, Cicerone, Rowland, Watson, Lovelock and Crutzen.

22 The European Environmental Office in Brussels criticised the Commission for endangering a potential convention by giving the integration of the EC a higher priority than the goal of protecting the ozone layer (Jachtenfuchs 1990: 264).

23 Incidentally, only in Germany is the term 'climatic catastrophe' (*Klimakatastrophe*) used. Elsewhere, talk is of 'global climate change' or even of the 'greenhouse effect'.

24 DMG and DPG, 1987. One insider told me the following about the catastrophe metaphor: 'That was so exaggerated that there was a strongly-worded protest from the Meteorological Society with the motto: If you make statements about climate, then be so kind as to ask us, the experts, because your Energy Working Group doesn't have a single climatologist in its own ranks . . . the meaning of the word catastrophe in fact says: Something is befalling us here that we couldn't have foreseen. But we do foresee it'.

25 'It was a great advantage that we had a lot of Americans at the hearings, German science was as a rule circumspect, unclear in its formulations. The scientists from the NASA programme, then Rowland, and the UNO representatives, they helped us a great deal. The Germans were reserved, E. or L. terribly so. The ones who were positive were C. and A., but of course they were purely scientific. Then there was Z., who in the end played the role that on the one hand he wanted to be a scientist, but he didn't want to pick a fight with the politicians' (GEPO 52). By early 1988, McFarland of Du Pont was already convinced of the seriousness of the ozone problem through his work on the OTP, and was probably not really anti-regulatory.

26 'German industry didn't succeed in stopping the preparations for Montreal again and ending them and changing the timetable. That had been set in about 1985 as far as the government was concerned, nothing more changed there. There's a point, the EK was one of them, where they didn't deal with the many scientific reports any more. They said: This question is decided for us, scientifically and politically, and decided in international environmental law, and the chemical industry will have to fall into line. That happened about 1985' (GEAD 12). A representative of German industry expressed his contempt for this process when he said: 'The Federal Republic wanted to be a good boy and play a leading political role' (GEIN 3).

6 The Montreal Protocol and after

1 '[B]efore that Würzburg meeting . . . the European governments said: We don't have the same results as the Americans, in terms of predictions. So they decided: Well, go to Würzburg and try to find out what's happening and run models as much as under the same conditions as you can. And we did that and we came up with very similar results. And then we went back and the negotiations went on on that basis. That played a key role in the preparation of the Montreal Protocol' (USAS 41).

2 A similar problem is described with the concept of 'incomplete contracting' in the neo-institutionalist literature (Williamson 1985). 'Incomplete contracting leads to problems of imperfect commitment. There is a strong temptation to renege on the original terms of the contract because what should be done in case of an unforeseen contingency is left unstated or ambiguous and thus open to interpretation. The problem is that the possibility of renegotiating deprives the original agreement of its credibility and prevents it from guiding behavior as intended' (Majone 1996b).

3 Essentially contained in Benedick (1991), Lang (1988) and Tolba (1998).

4 At the mention of the official version, claiming that the ozone hole played no role at Montreal, this interviewee replied 'Bullshit!'.

5 A reason to be very nervous, as one NASA researcher explained: 'There was great secrecy about the second expedition, people were not allowed to go down there . . . I wasn't even allowed to go down there. This non-open way was changed subsequently. But then there was very much nervousness about the whole thing' (USAS 15).

6 In implementing the Montreal Protocol, the EC employed the rare legal mechanism of a regulation (EWG 594/91), in order to leave no leeway in terms of implementation at the national level (Salter 1996: 2/22–2/23; Jachtenfuchs 1990: 269). The same groups of substances are regulated as in the Montreal Protocol, but with shorter deadlines (Bundesregierung 1994: 13).

7 Elster (1989: 80–1) points out that there are structurally weak and strong negotiating positions, which depend on different evaluations of the results of negotiations and on the willingness of the parties to take risks. In the first case there is a 'Matthew effect', since a wealthy man in negotiations with a poor one can say: 'Take it or leave it.' When there is a discrepancy in levels of the willingness to take risks the same asymmetry appears: 'Whatever the source of risk aversion, it is usually a handicap in bargaining.'

8 However, a new conflict arose between the USA and the 'rest of the world' over the creation of an aid fund for developing countries; see below.

9 'U.S. Will Oppose Aid in Ozone Plan' (*New York Times*, 9 May 1990: A 24).
10 Twelve Republican senators wrote in a letter to Bush: 'There has rarely been, we believe, a better example of "penny-wise, pound-foolish"' (*Washington Post*, 10 May 1990).
11 'U.S. is Assailed at Geneva Talks for Backing out of Ozone Plan', read the front page of the *New York Times*, 10 May 1990.
12 'Regarding the partially halogenated, that is, the HCFCs, it was becoming clear that there would be new regulations and so they've hardly caught on in Germany. In the US they've gone more in the direction of [these] substitute materials in the chemical industry. So the EPA was always totally amazed by our position on HCFCs as well, they always took us for odd foreigners, although developments showed that we weren't so far wrong. At the time when Montreal happened, certainly no one would have ever bet that someday HCFCs would be regulated, now we've done it, even if it still says 2030 or whatever on it' (GEAD 24).
13 A scientist from the National Science Foundation says of the ozone hole in 1989, which had grown relative to that of 1988: 'It's terrifying. If these ozone holes keep growing like this, they'll eventually eat the world' (*New York Times*, 23 September 1989).
14 Goodell draws a pyramid model of the visibility of scientists. At the bottom are scientists who are rarely heard from (either because they are not interested in publicity or the media have no interest in them). Above them is a layer of scientists whose work becomes newsworthy for a short time. Closer to the top there are scientists who are often quoted in the media as specialists on a specific topic. Further to the top are scientists who espouse their opinions about general problems and the priorities of science policy. On top are those scientists who are motivated, who can be quoted, who are colourful, credible and accessible to become celebrities. They often have very strong opinions and take part on one side of a scientific controversy (Goodell 1987: 593–4).
15 For Shils there are no institutional rules for scientists who dare to leave the core of science: 'There is as yet no sound tradition such as exists at the heart of science to guide action in these activities that are at the periphery of science itself but are of the greatest importance to society' (Shils 1987: 202).
16 Cf. the following statement of an American scientist: 'Science does not exist in a vacuum. There is an old Polish saying "The guy would starve to death unless a pigeon flew into his mouth"– If you just stand on the hillside with your mouth open, in science, you may be producing the greatest amount of work, but you've got to sell it to show that your work is worth funding. When you have limited resources, you have to show that your science is better than anybody else's. Sometimes that is exaggerated' (USAS 17).
17 *Der Spiegel* published the cover story 'Ozone hole over Europe' on 10 February 1992.
18 The scientists I questioned were critical on this point: 'This was a little bit overstated .... They were operating out of Maine and Bush was there on holiday, and they were telling that you could have [an ozone hole]. That was premature, they were too excited that they found some of the culprits in between reaction species which confirmed the view of ozone destruction' (UNAS 2). 'That had a big impact in the community, it was said we spoke too soon, I was not involved in it, but it was a mistake' (USAS 17). 'The decision of Bush to ban came immediately after Al Gore went on television and talked about a hole in ozone over Kennebunkport which might get George Bush's attention' (USAS 30).

7 Lessons

1 This correlation between a country's active policy and high expectations in the relevant public sphere also appears to apply, albeit in reverse, to the climate conferences at Rio (1992) and Berlin (1995), where the German press, almost 'proportional' to the high level of commitment shown by the German government, produced a level of expectation that was not matched in the American press. 'The Berlin summit was barely mentioned in the US papers' (personal communication from Allan Mazur).

2 Undoubtedly, factual errors occur, for example, after the Nobel Prize was awarded to Crutzen, Molina and Rowland, *Der Spiegel* ((42)1995: 272) names these three as the discoverers of the ozone hole. Upon enquiry, it became clear that no one at the magazine had ever heard of the British Antarctic team and Joe Farman (telephone conversation between the author and the science editor, 2 November 1995).

3 In contrast to the NRDC, which in the 1970s already functioned in the USA as part of the advocacy alliance. It was a very new organisation, one of the NGOs created in reaction to Earth Day 1970.

4 I am giving a somewhat stylised account of Haas's position. He has taken pains to avoid a monocausal explanation of the Montreal Protocol himself (see Haas 1993). But his approach has been taken up by others as the epistemic community approach and thus become very influential – which is why I feel legitimised in stylising him the way I do.

5 If one takes the overlap between model predictions and measurements as an indicator of the underlying scientific understanding, it is said to be somewhere between 60 per cent and 90 per cent in the case of polar ozone, whereas in the case of mid-latitudes the fit is only 50 per cent (as various interviewees told me).

6 Likewise, Ian Rowlands (not to be confused with F. S. Rowland) observes that the politicians 'signed the agreement in September 1987, while scientists did not publish their report until March 1988' (Rowlands 1995: 30).

7 'Spencer and Durkheim, of course, would have been optimistic, since they postulated that increased differentiation was followed by the rise of integrating agencies that brought together the parts. But this aspect of their theories seems to be inaccurate, and we have differentiation without integration and seem quite possibly fated to have it for a very long time to come' (Collins 1986: 1340).

8 Although Merton was aware of this in a different context, see Merton (1973b).

9 Reportedly, the physicist and Nobel laureate Steven Weinberg feels that holding on to a particular world-view leads to making discoveries: 'It's for the best if physicists don't believe in the anthropic principle, because otherwise they're not motivated to look for a unifying theory, and if they weren't looking for that, they certainly wouldn't find anything' (cit. in Brockman 1995).

10 Cf. Beck's dismissive judgement on the sciences: '[The sciences,] as they are constituted, . . . are *entirely incapable* of reacting adequately to civilizational risks, since they are prominently involved in the origin and growth of those very risks. Instead, . . . [they] become the *legitimating patrons* of a global industrial pollution and contamination of air, water, foodstuffs, etc., as well as the related generalized sickness and death of plants, animals and people' (Beck 1992: 59, emphasis in original; cf. Mills 1961, 1963; and Restivo 1988).

11 As cabinet minister Schmidbauer lucidly observed in German parliament: 'By these means the scientific findings will increase; the drama of the situation can be made clearer and so, naturally, the pressure to take political measures can be intensified as well' (Deutscher Bundestag 1988: 6436).

8 Epilogue: the example of climate

1 This section draws upon a longer version, see Grundmann (forthcoming).

2 After the departure of Bert Bolin as chair of the Working Group, Bob Watson was proposed as his successor, 'since he was the one who worked the magic before' (Ralph Cicerone, cit. in Litfin 1994: 207).

3 For example, ICI stated that doing without CFCs 'would not only mean changing a lifestyle, it would affect health, safety and economies, and not only in the western world; third world countries would be affected as well' (cit. in Purvis *et al.* 1997).

4 Wuebbles and Rosenberg (1998: 66) indicate that there was a great deal of controversy over the GWP indicator. As a result of the carbon cycle, the indicator is time-dependent and is typically applied to three temporal horizons: to periods of 20, 100 and 500 years.

Bibliography

Adam, B. (2000) The temporal gaze: the challenge for social theory in the context of GM food, *The British Journal of Sociology* 51(1): 125–42.

Adams, J. (1995) *Risk*, London: UCL Press.

Adler, E. and Haas, P. M. (1992) Conclusion: epistemic communities, world order, and the creation of a reflective research program, *International Organization* 46: 367–90.

Alexander, J. (ed.) (1985) *Neofunctionalism*, Beverly Hills, CA: Sage.

Alexander, J. and Colomy, P. (eds) (1990) *Differentiation Theory and Social Change, Comparative and Historical Perspectives*, New York: Columbia University Press.

AFEAS (Alternative Fluorocarbons Environmental Acceptability Study) (1992) *Production, Sales and Atmospheric Release of Fluorocarbons Through 1992*, Washington, DC: AFEAS.

American Geophysical Union (1989) The airborne Antarctic ozone experiment (AAOE), *Journal of Geophysical Research* 94, Special issue.

Anderson, J. G., Brune, W. H. and Proffitt, M. H. (1989) Ozone destruction by chlorine radicals within the Antarctic vortex: the spatial and temporal evolution of $ClO-O_3$ anticorrelation based on in situ ER-2 data, *Journal of Geophysical Research* 94: 11465–79.

Arthur, W. B. (1988) Self-reinforcing mechanisms in economics, in Philip W. Anderson (ed.) *The Economy as an Evolving Complex System*, Redwood City, CA: Addison-Wesley, 9–31.

Arthur, W. B. (1990) Positive feedbacks in the economy, *Scientific American* 262(2): 92–9.

Arthur, W. B., Ermoliev, Y. M. and Kaniovski, Y. M. (1987) Path-dependent processes and the emergence of macro-structure, *European Journal of Operational Research* 30: 294–303.

Ausubel, J. H. (1989) Protecting the ozone layer: a perspective from industry, in J. H. Ausubel and H. E. Sladovitch (eds) *Technology and Environment*, Washington, DC: National Academy Press, 70–94.

Axelrod, R. (1984) *The Evolution of Cooperation*, New York: Basic Books.

Axelrod, R. and Keohane, R.O. (1986) Achieving cooperation under anarchy: strategies and institutions, in K. A. Oye (ed.) *Cooperation Under Anarchy*, Princeton, NJ: Princeton University Press, 226–54.

Bacharach, P. and Baratz, M.S. (1970) *Power and Poverty. Theory and Practice*, New York: Oxford University Press.

Barnes, B. (1990) Sociological theories of scientific knowledge, in R. C. Olby *et al.* (eds) *Companion to the History of Modern Science*, London: Routledge, 60–73.

Barnes, B. and Dolby, R. G. A. (1970) The scientific ethos: a deviant viewpoint, *European Journal of Sociology* 11: 3–25.

Bastian, C. L. (1982) The formulation of Federal policy, in F. A. Bower and R. B. Ward (eds) *Stratospheric Ozone and Man*, Boca Raton: CRC Press, 164–200.

Beck, U. (1992) *Risk Society*, London: Sage.

Beck, U. (1996) Risk society and the provident state, in S. Lash, B. Szerzynsky and B. Wynne (eds) *Risk, Environment and Modernity. Towards a New Ecology*, London: Sage, 27–43.

Beck, U., Giddens, A. and Lash, S. (eds) (1995) *Reflexive Modernization: Politics, Tradition and Aesthetics in the Modern Social Order*, Stanford, CA: Stanford University Press.

Beckerman, W. (1995) *Small is Stuid – Blowing the Whistle on the Greens*. London: Duckworth.

Ben-David, J. (1960) Roles and innovations in medicine, *American Journal of Sociology* 65(6): 557–68.

Ben-David, J. (1991a) [1960] Scientific productivity and academic organization in nineteenth-century medicine, in J. Ben-David (ed.) *Scientific Growth, Essays on the Social Organization and Ethos of Science*, Berkeley, CA: University of California Press, 103–24.

Ben-David, J. (1991b) 'Norms of Science' and the sociological interpretation of scientific behavior, posthumous publication, in J. Ben-David (ed.) *Scientific Growth, Essays on the Social Organization and Ethos of Science*, Berkeley, CA: University of California Press, 469–84.

Benedick, Richard E. (1991) *Ozone Diplomacy, New Directions in Safeguarding the Planet*, Cambridge, MA: Harvard University Press.

Benz, A. (1994) *Kooperative Verwaltung, Funktionen, Voraussetzungen und Folgen*, Baden-Baden: Nomos.

Benz, A. (1995) Verhandlungssysteme und Mehrebenenverflechtung im kooperativen Staat, in W. Seibel and A. Benz (eds) *Regierungssystem und Verwaltungspolitik, Beiträge zu Ehren von Thomas Ellwein*, Opladen: Westdeutscher Verlag, 83–102.

Berger, S. (ed.) (1981) *Organizing Interests in Western Europe, Pluralism, Corporatism, and the Transformation of Politics*, Cambridge: Cambridge University Press.

Berry, Jeffrey N. (1977) *Lobbying for the People, The Political Behavior of Public Interest Groups*, Princeton, NJ: Princeton University Press.

Betsill, M. and Pielke, R. Jr. (1998) Blurring the boundaries: domestic and international ozone politics and lessons for climate change, *International Environmental Affairs* 10: 147–72.

Beutler, B. *et al.* (1993) *Die Europäische Union*, Baden-Baden: Nomos.

Biermann, F. (1997) Financing environmental policies in the south: experiences from the multilateral ozone fund, *International Environmental Affairs* 9: 179–218.

Black, M. (1961) *Models and Metaphors, Studies in Language and Philosophy*, Ithaca and London: Cornell University Press.

Bloor, D. (1976) *Knowledge and Social Imagery*, London: Routledge.

Blumenthal, D. *et al.* (1997) Withholding research results in academic life science – evidence from a national survey of faculty, *Journal of the American Medical Association* 277: 1224–8.

Bodansky, D. (1994) Prologue to the climate change convention, in I. Mintzer and J. A. Leonard (eds) *Negotiating Climate Change: The Inside Story of the Rio Convention*, Cambridge: Cambridge University Press, 45–74.

Boehmer-Christiansen, S. (1994a) Global climate protection policy: the limits of scientific advice (Part 1), *Global Environmental Change* 4: 140–59.

Boehmer-Christiansen, S. (1994b) Global climate protection policy: the limits of scientific advice (Part 2), *Global Environmental Change* 4: 185–200.

Boffey, P. (1975) *The Brain Bank of America*, New York: McGraw-Hill.

Böhme, G. and Schramm, E. (eds) (1985) *Soziale Naturwissenschaft: Wege zu einer Erweiterung der Ökologie*, Frankfurt am Main: Fischer.

Bohne, E. (1984) Informales Verwaltungs und Regierungshandeln als Instrument des Umweltschutzes, *Verwaltungs-Archiv* 75: 343–73.

Bohne, E. (1990) Recent trends in informal environmental conflict resolution, in

W. Hoffmann-Riem and E. Schmidt-Aßmann (eds) *Konfliktbewältigung durch Verhandlungen*, Baden-Baden: Nomos, 217–30.

Bonß, W. (1996) Die Rückkehr der Unischerheit. Zur gesellschaftstheoretischen Bedeutung des Risikobegriffs, in G. Banse (ed.) *Risikoforschung zwischen Disziplinarität und Interdisziplinarität*, Berlin: Sigma, 165–84.

Boventer, H. (1993) Ohnmacht der Medien. Die Kapitulation der Medien vor der Wirklichkeit. *Aus Politik und Zeitgeschichte. Beilage zum Parlament* B40: 27–35.

Brack, D. (1996) *International Trade and the Montreal Protocol*, London: Royal Institute of International Affairs.

Brand, K.-W. (1995) Der ökologische Diskurs. Wer bestimmt Themen, Formen und Entwicklung der öffentlichen Umweltdebatte?, in Gerhard de Haan (ed.) *Umweltbewußtsein und Massenmedien*, Berlin: Akademie, 47–62.

Brasseur, G. P. (1988) Group report: changes in atmospheric ozone, in F. S. Rowland and I. S. Isaksen (eds) *The Changing Atmosphere*, Chichester: John Wiley & Sons, 235–56.

Braun, D. (1993) Zur Steuerbarkeit funktionaler Teilsysteme: Akteurtheoretische Sichtweisen funktionaler Differenzierung moderner Gesellschaften, in A. Héritier (ed.) *Policy-Analyse, Kritik und Neuorientierung. Politische Vierteljahresschrift, Sonderheft* 24: 199–222.

Breitmeier, H. (1996) *Wie entstehen globale Umweltregime?* Opladen: Leske & Budrich.

Breitmeier, H. (1997) International organisations and the creation of international regimes, in O. R. Young (ed.) *Global Governance. Drawing Insights from Environmental Experience*, Cambridge, MA: The MIT Press.

Broad, W. and Wade, N. (1982) *Betrayers of the Truth*, New York: Simon and Schuster.

Brockman, J. (1995) *The Third Culture*, New York: Simon and Schuster.

Brodeur, P. (1975) Annals of chemistry, inert, *The New Yorker*, 7 April: 47–56.

Brodeur, P. (1986) Annals of chemistry: in the face of doubt, *The New Yorker*, 9 June: 70–86.

Brooks, H. (1982) Stratospheric ozone, the scientific community and public policy, in Frank A. Bower and Richard B. Ward (eds) *Stratospheric Ozone and Man*, Boca Raton: CRC Press, 201–16.

Brooks, H. and Cooper, C. L. (eds) (1987) *Science for Public Policy*, Oxford: Pergamon Press.

Brown, G. (1996) Environmental science under siege: fringe science and the 104th Congress. A Report by Representative George E. Brown, Ranking Democratic Member to the Democratic Caucus of the Committee on Science US House of Representatives, 23 October.

Bucchi, M. (1998) *Science and the Media. Alternative Routes in Scientific Communication*, London: Routledge.

Bundesregierung (1986) Antwort auf die Anfrage der Abgeordneten Frau Hönes, Schmidt (Hamburg-Neustadt) und der Fraktion Die Grünen, Bonn: Bundestagsdrucksache 10/6724.

Bundesregierung (1987) Antwort der Bundesregierung auf die Anfrage der Fraktion Die Grünen, Bonn: Bundestagsdrucksache 10/5400.

Bundesregierung (1994) Dritter Bericht der Bundesregierung an den Deutschen Bundestag über Maßnahmen zum Schutz der Ozonschicht, Bonn: Drucksache 12/8555.

Burhenne, W. E. and Kehrhahn, J. (1981) Neue Formen parlamentarischer Zusammenarbeit, in H. J. Vogel *et al.* (eds) *Die Freiheit des Anderen: Festschrift für Martin Hirsch*, Baden-Baden: Nomos.

Burningham, K. and Cooper, G. (1999) Being constructive: social constructionism and the environment, *Sociology* 33: 297–316.

Cagin, S. and Dray, P. (1993) *Between Earth and Sky, How CFCs Changed Our World and Endangered the Ozone Layer*, New York: Pantheon Books.

Caldwell, L. K. (1984) The world environment: reversing US policy commitments, in N. J. Vig and M. E. Kraft (eds) *Environmental Policy in the 1980s*, Washington, DC: CQ Press, 319–38.

Callon, M. (1987) Society in the making: the study of technology as a tool for sociological analysis, in W. E. Bijker *et al.* (eds) *The Social Construction of Technological Systems, New Directions in the Sociology and History of Technology*, Cambridge, MA: The MIT Press, 83–103.

Callon, M. (1995) Four models for the dynamics of science, in Sheila Jasanoff *et al.* (eds) *Handbook of Science and Technology Studies*, London: Sage, 29–63.

Carmody, K. (1995) Environmental journalism in an age of backlash, *Columbia Journalism Review*, May–June: 40–5.

Carson, R. (1962) *Silent Spring*, Boston: Houghton Mifflin.

Chubachi, S. (1984) Preliminary result of ozone observations at Syowa Station from February 1982 to January 1983. *Memoirs of National Institute of Polar Research, Proceedings of the Sixth Symposium on Polar Meteorology and Glaciology*, Special Issue No 34, Tokyo: National Institute of Polar Research, 13–19.

Chubin, D. E. and Hackett, E. J. (1990) *Peerless Science, Peer Review and US Science Policy*, Albany, NY: SUNY.

CIAP (Climatic Impact Assessment Program) (1973) *Impact of High-Flying Aircraft on the Stratosphere*, Washington, DC: Department of Transportation.

Clausen, L. and Dombrowsky, W. R. (1984) Warnpraxis und Warnlogik, *Zeitschrift für Soziologie* 13: 293–307.

Cohen, M. D., March, J. and Olsen, J. P. (1972) A garbage can model of organizational choice, *Administrative Science Quarterly* 17: 1–19.

Coleman, J. S. (1982) *The Asymmetric Society*, Syracuse, NY: Syracuse University Press.

Coleman, J. S. (1985) Responsibility in corporate action: a sociologist's view, in K. Hopt and G. Teubner (eds) *Corporate Governance and Directors' Liabilities: Legal, Economic and Sociological Analyses on Corporate Social Responsibility*, Berlin: de Gruyter, 69–91.

Collingridge, D. and Reeve, C. (1986) *Science Speaks to Power: The Role of Experts in Policy*, London: Pinter.

Collins, H. M. (1985) *Changing Order*, London: Sage.

Collins, R. (1986) Is 1980s sociology in the doldrums?, *American Journal of Sociology* 91: 1336–55.

Cooper, R. N. (1989) International cooperation in public health as a prologue to macro-economic cooperation, in R. Cooper *et al.* (eds) *Can Nations Agree? Issues in International Economic Cooperation*, Washington, DC: The Brookings Institution, 178–254.

Cornes, R. and Sandler, T. (1994) Are public goods myths?, *Journal of Theoretical Politics* 6: 369–85.

Crane, D. (1965) Scientists at major and minor universities, *American Sociological Review* 30: 699–714.

Crane, D. (1972) *Invisible Colleges, Diffusion of Knowledge in Scientific Communities*, Chicago, IL: The University of Chicago Press.

Crawford, M. (1987) Ozone plan: tough bargaining ahead, *Science*, 4 September: 1099.

Crutzen, P. J. (1970) The influence of nitrogen oxides on the atmospheric ozone content, *Quarterly Journal of the Royal Meteorological Society* 96: 320–5.

Crutzen, P. J. (1989) Das Ozonloch hat fast eine Revolution in Gang gesetzt (Interview), *Die Welt*, 2 October: 7.

Crutzen, P. J. (1996) Mein Leben mit O_3, NO_x und anderen YZO_x-Verbindungen (Nobel-Vortrag), *Angewandte Chemie* 108: 1878–98.

Crutzen, P. J. and Arnold, F. (1986) Nitric-acid cloud formation in the cold Antarctic stratosphere – a major cause for the springtime ozone hole, *Nature* 324: 651–5.

Cunnold *et al.* (1983) The atmospheric lifetime experiment. Lifetime methodology and application to three years of $CFCl_3$ data, *Journal of Geophysical Research* 88: 8379–400.

Dasgupta, P. and David, P. A. (1994) Toward a new economics of science, *Research Policy* 23: 487–521.

David, P. A. (1985) Clio and the economics of QWERTY, *American Economic Review* 75: 332–7.

de Haan, G. (1995) Ökologische Kommunikation. Der Stand der Debatte, in G. de Haan (ed.) *Umweltbewußtsein und Massenmedien*, Berlin: Akademie, 17–34.

DeLeon, P. (1994) The policy sciences redux: new roads to post-positivism, *Policy Studies Journal* 22: 176–84.

Deutsche Forschungsgemeinschaft (DFG) (1985) *Sonderforschungsbereiche 1969–1984*, Weinheim: VCH.

Deutsche Meteorologische Gesellschaft und Deutsche Physikalische Gesellschaft (1987) *Warnung vor drohenden weltweiten Klimaänderungen durch den Menschen*, Bad Honnef.

Deutscher Bundestag (1988) Plenarprotokoll der 94. Sitzung des Deutschen Bundestags, 11. Wahlperiode, 22. September.

Diamond, J. (1997) Kinship with the stars, *Discover* 16: 44.

Dickens, P. (1996) *Reconstructing Nature*, London: Routledge.

Dickman, S. (1987) US call to end CFC emissions, *Nature* 325: 748.

Doniger, D. (1988) Politics of the ozone layer, *Issues in Science and Technology*, Spring: 86–92.

Dotto, L. and Schiff, H. (1978) *The Ozone War*, New York: Doubleday.

Douglas, M. (1982) Cultural bias, in Mary Douglas *In the Active Voice*, London: Routledge and Kegan Paul, 183–254.

Douglas, M. (1986) *How Institutions Think*, Syracuse, NY: Syracuse University Press.

Douglas, M. (1988) A typology of cultures, in H. -J. Hoffmann-Nowottny and W. Zapf (eds) *Kultur und Gesellschaft, Verhandlungen des 24, Deutschen Soziologentags und des 8, Kongresses der Schweizerischen Gesellschaft für Soziologie in Zürich 1988*, Frankfurt am Main: Campus, 85–97.

Douglas, M. and Wildavsky, A. (1982) *Risk and Culture*, Berkeley, CA: University of California Press.

Dowie, M. (1995) *Losing Ground. American Environmentalism at the Close of the Twentieth Century*, Cambridge, MA: The MIT Press.

Downie, D. L. (1995) Road map or false trail? Evaluating the 'precedence' of the ozone regime as a model and strategy for global climate change, *International Environmental Affairs* 7: 321–45.

Downs, A. (1972) Up and down with ecology – the 'Issue Attention Cycle', *The Public Interest* 28: 38–50.

Downs, G. W., Rocke, D. M. and Siverson, R. M. (1986) Arms races and cooperation, in K. A. Oye (ed.) *Cooperation Under Anarchy*, Princeton, NJ: Princeton University Press, 80–117.

Dresselhaus, M. S. (1995) National science policy: the American experience, in I. Asher *et al.* (eds) *Strategies for the National Support of Basic Research: An International Comparison*, Jerusalem: The Israel Academy of Sciences and Humanities, 65–72.

Dryzek, J. S. (1997) *The Politics of the Earth, Environmental Discourses*, Oxford: Oxford University Press.

Dudek, D. J., Leblanc, A. and Sewall, K. (1990) Cutting the cost of environmental policy: lessons from business response to CFC regulation, *Ambio* 19: 324–8.

Elfenbein, S. (1996) *Die New York Times, Mythos und Macht eines Mediums*, Frankfurt am Main: Fischer.

Elias, N. and Scotson, J. (1965) *The Established and the Outsiders*, London: Frank Cass.

Elsaesser, H. W. (1978) Ozone destruction by catalysis: credibility of the threat, *Atmospheric Environment* 12: 1849–56.

Elsaesser, H. W. (1994) The unheard arguments: a rational view on stratospheric ozone, *21st Century*, Fall: 38–45.

Elster, J. (1978) *Logic and Society, Contradictions and Possible Worlds*, Chichester: Wiley.

Elster, J. (1979) Risk, uncertainty and nuclear power, *Social Science Information* 18: 371–400.

Elster, J. (1989) *The Cement of Society, A Study of Social Order*, Cambridge: Cambridge University Press.

Elster, J. (1990) When rationality fails, in K. Cook and M. Levi (eds) *The Limits of Rationality*, Chicago, IL: University of Chicago Press, 19–51.

Elzinga, A. (1995) Shaping worldwide consensus: the orchestration of global climate change research, in A. Elzinga and C. Lundström (eds) *Internationalism in Science*, London: Taylor and Graham.

Enquetekommission des 11. Deutschen Bundestages 'Vorsorge zum Schutz der Erdatmosphäre' (1990) *Schutz der Erdatmosphäre, Eine internationale Herausforderung*, 2nd edn, Bonn/Karlsruhe: Economica/C.F. Müller.

EPA (Environmental Protection Agency) (1987) *Assessing the Risk of Trace Gases That Can Modify the Stratosphere*, Washington, DC: EPA.

Evers, A. and Nowotny, H. (1989) Über den Umgang mit Unsicherheit, in U. Beck and W. Bonß (eds) *Weder Sozialtechnologie noch Aufklärung? Analysen zur Verwendung sozialwissenschaftlichen Wissens*, Frankfurt am Main: Suhrkamp, 355–83.

Ezrahi, Y. (1990) *The Descent of Icarus: Science and the Transformation of Contemporary Democracy*, Cambridge, MA: Harvard University Press.

Falkner, R. (1998) The multilateral ozone fund of the Montreal Protocol, *Global Environmental Change* 8: 171–5.

Farman, J. C., Gardiner, B. G. and Shanklin, J. D. (1985) Large losses of total ozone in Antarctica reveal seasonal ClO_x/NO_x interaction, *Nature* 315: 207–10.

Festinger, L. (1957) *A Theory of Cognitive Dissonance*, Stanford, CA: Stanford University Press.

Fischer, F. (1998) Beyond empiricism: policy inquiry in postpositivist perspective, *Policy Studies Journal* 26(1): 129–46.

Fischer, F. and Forester, J. (eds) (1993) *The Argumentative Turn in Policy Analysis and Planning*, Durham, NC and London: Duke University Press.

Fischhoff, B. *et al.* (1981) *Acceptable Risk*, Cambridge: Cambridge University Press.

Fish, S. (1994) Being interdisciplinary is so very hard to do, in S. Fish (ed.) *There is No Such Thing as Free Speech*, New York: Oxford University Press.

Fleig, A. J. *et al.* (1986) Seven years of total ozone from the TOMS instrument – a report on data quality, *Geophysical Research Letters* 13: 1355– 8.

Fleig, A. J., Bhartia, P. K. and Silberstein, D. S. (1986) An assessment of the long-term drift in SBUV total ozone data, based on comparison with the Dobson Network, *Geophysical Research Letters* 13: 1359–62.

Føllesdal, D. (1979) Some ethical aspects of recombinant DNA research, *Social Science Information* 18: 401–19.

Forester, J. (1993) Learning from practice stories: the priority of practical judgement, in F. Fischer and J. Forester (eds) *The Argumentative Turn in Policy Analysis and Planning*, Durham, NC: Duke University Press, 186–209.

Friedland, J. (1999) Dr Evil and his Moneyman: we're all flocking to get into this one, *Guardian*, 17 February: 18.

Friedman, R. S. and Friedman, R. C. (1990) American science, academic organisation, and interdisciplinary research, in P. Birnbaum-More, F. A. Rossini and D. R. Baldwin *International Research Management. Studies in Interdisciplinary Methods from Business, Government and Academia*, New York: Oxford University Press.

Fuchs, S. (1993a) Positivism is the organizational myth of science, *Perspectives on Science* 1: 1–23.

Fuchs, S. (1993b) A sociological theory of scientific change, *Social Forces* 71: 933–53.

Fülgraff, G. (1994) Perspektiven für die Arbeit einer wissenschaftlichen Umweltbehörde, in Umweltbundesamt (ed.) *Wissenschaften im ökologischen Wandel*, Berlin: Umweltbundesamt.

Funtowicz, S. O. and Ravetz, J. R. (1992) Three types of risk assessment and the emergence of post-normal science, in S. Krimsky and D. Golding (eds) *Social Theories of Risk*, Westport: Praeger, 251–73.

Gadamer, H. -G. (1960) *Wahrheit und Methode*, Tübingen: Mohr (*Truth and Method*, translated from the German; translation edited by Garrett Barden and John Cumming, London: Sheed and Ward [1975]).

Galison, P. (1987) *How Experiments End*, Chicago, IL: University of Chicago Press.

Gamson, W. A. and Modigliani, A. (1989) Media discourse and public opinion on nuclear power: a constructionist approach, *American Journal of Sociology* 95: 1–37.

Garfield, E. (1988) Ozone layer depletion: its consequences, the causal debate, and international cooperation, *Current Contents* 20(6): 3–13.

Garfield, E. (1992) Robert T. Watson of NASA receives NAS award for scientific reviewing of stratospheric ozone dynamics, *Current Contents* 24(17): 5–10.

Garrett, G. and Weingast, B. (1993) Ideas, interests and institutions: constructing the European Community's internal market, in J. Goldstein and R. O. Keohane (eds) *Ideas and Foreign Policy, Beliefs, Institutions, and Political Change*, Ithaca: Cornell University Press, 173–206.

Gehring, T. (1994) *Dynamic International Regimes. Institutions for International Environmental Governance*, Frankfurt am Main: Peter Lang.

Gelbspan, R. (1997) *The Heat is on: The Climate Crisis, the Cover-Up, the Prescription*, Reading, MA: Addison-Wesley.

Genscher, Hans-Dietrich (1995) *Erinnerungen*, Berlin: Siedler.

Gerhards, J. and Neidhardt, F. (1990) *Strukturen und Funktionen moderner Öffentlichkeit*, WZB Discussion Paper FS III 91–108, Berlin: Wissenschaftszentrum Berlin für Sozialforschung (WZB).

Gibbons, M. *et al.* (1994) *The New Production of Knowledge, The Dynamics of Science and Research in Contemporary Societies*, London: Sage.

Giddens, A. (1990) *The Consequences of Modernity*, Cambridge: Polity.

Gieryn, T. F. (1995) Boundaries of science, in Sheila Jasanoff *et al.* (eds) *Handbook of Science and Technology Studies*, London, Sage, 393–443.

Glas, J. P. (1989) Protecting the ozone layer: a perspective from industry, in J. Ausubel and H. Sladovitch (eds) *Technology and Environment*, Washington, DC: National Academy Press, 137–55.

Godwin, R.K./Robert C. Mitchell (1984) The impact of direct mail on political organizations, *Social Science Quarterly* 65: 829–39.

Goffman, E. (1975) *Frame Analysis, an Essay on The Organization of Experience*, Harmondsworth: Penguin.

Goldstein, J. and Keohane, R. (1993) Ideas and foreign policy: an analytical framework, in J. Goldstein and R. Keohane (eds) *Ideas and Foreign Policy, Beliefs, Institutions, and Political Change*, Ithaca: Cornell University Press, 3–30.

Goodell, R. (1977) *The Visible Scientists*, Boston, MA: Little Brown.

Goodell, R. (1987) The role of the mass media in scientific controversy, in H. Engelhardt and A. L. Kaplan (eds) *Scientific Controversies, Case Studies in the Resolution and Closure of Disputes in Science and Technology*, Cambridge: Cambridge University Press, 585– 97.

Goss-Levi, B. (1988) Ozone depletion at the poles: the hole story emerges, *Physics Today* 41(7): 17–21.

Graedel, T. E. and Crutzen, P. J. (1995) *Atmosphere, Climate, and Change*, New York: Scientific American Library.

Granovetter, M. (1978) Threshold models of collective behavior, *American Journal of Sociology* 83: 1420– 43.

Grundmann, R. (1991) *Marxism and Ecology*, Oxford: Oxford University Press.

Grundmann, R. (1998) Technik als Problem für die Systemtheorie [Technology as a problem for systems theory], *Swiss Journal of Sociology* 24(2): 327–46.

Grundmann, R. (1999) Sociology and nature: social action in context – review of Raymond Murphy, Sociology and Nature, *Canadian Journal of Sociology* 24(2): 317–20.

Grundmann, R. (forthcoming) Ozone and Climate: have we learnt the right lessons? Ms., submitted to *Climatic Change*.

Haas, E. (1990) *When Knowledge is Power*, Berkeley, CA: University of California Press.

Haas, P. M. (1992a) Introduction: epistemic communities and international policy coordination, *International Organization* 46: 1–35.

Haas, P. M. (1992b) Banning chlorofluorocarbons: epistemic community efforts to protect stratospheric ozone, *International Organization* 46: 187–224.

Haas, P. M. (1993) Stratospheric ozone: regime formation in stages, in O. Young and G. Osherenko (eds) *Polar Politics, Creating International Environmental Regimes*, Ithaca: Cornell University Press, 152–85.

Haas, P. M. *et al.* (eds) (1993) *Institutions for the Earth*, Cambridge, MA: The MIT Press.

Haggard, S. and Simmons, B. A. (1987) Theories of international regimes, *International Organization* 41: 491–517.

Hagstrom, W. O. (1965) *The Scientific Community*, New York: Basic Books.

Hahn, R. W. (1990) The political economy of environmental regulation: towards a unifying framework, *Public Choice* 65: 21–45.

Hahn, R. W. and Richards, K. R. (1989) The internationalization of environmental regulation, *Harvard International Law Review* 30: 421–46.

Haigh, N. (1992) The European Community and international environmental policy, in A. Hurrell and B. Kingsbury (eds) *The International Politics of the Environment*, Oxford: Oxford University Press, 228– 49.

Hajer, M. A. (1993) Discourse coalitions and the institutionalization of practice: the case of acid rain in Great Britain, in F. Fischer and J. Forester (eds) *The Argumentative Turn in Policy Analysis and Planning*, Durham, NC: Duke University Press, 43–76.

Hajer, M. A. (1995) *The Politics of Environmental Discourse: Ecological Modernisation and the Policy Process*, Oxford: Oxford University Press.

Hall, J. A. (1993) Ideas and the social sciences, in J. Goldstein and R. O. Keohane (eds) *Ideas and Foreign Policy, Beliefs, Institutions, and Political Change*, Ithaca: Cornell University Press, 31–54.

Hall, P. (ed.) (1989) *The Political Power of Economic Ideas, Keynesianism Across Nations*, Princeton, NJ: Princeton University Press.

Hall, P. and Taylor, R. (1996) Political science and the three new institutionalisms, *Political Studies* 44: 936–57.

Hannigan, J. (1995) *Environmental Sociology, A Social Constructionist Perspective*, London: Routledge.

Hardin, G. (1978) Political requirements for preserving our commons heritage, in H. P. Brokaw (ed.) *Wildlife and America*, Washington, DC: Council on Environmental Quality, 310–17.

Hardin, G. (1980) The tragedy of the *commons*, in Herman E. Daly (ed.) *Economics, Ecology, Ethics*, San Francisco, CA: W. H. Freeman.

Hardin, R. (1982) *Collective Action*, Baltimore, MD: Johns Hopkins University Press.

Hart, D. M. and Victor, D. G. (1993) Scientific elites and the making of US policy for climate change research 1957–74, *Social Studies of Science* 23: 643–80.

Hartkopf, G. and Bohne, E. (1983) *Umweltpolitik*, Vol. 1, Opladen: Westdeutscher Verlag.

Harvey, D. (1998) University, Inc., *Atlantic Monthly* (October). http://www.theatlantic.com/issues/98oct/ruins.htm

Hasselmann, K. (1998) Cooperative and non-cooperative multi-actor strategies of optimizing greenhouse gas emissions, in H. von Storch *et al.* (eds) *Anthropogenic Climate Change*, Berlin: Springer, 219–69.

Hays, S. P. (1987) *Beauty, Health and Permanence, Environmental Politics in the United States (1955–1985)*, Cambridge: Cambridge University Press.

Heclo, H. (1978) Issue networks and the executive establishments, in A. D. King (ed.) *The New American Political System*, Washington, DC: American Enterprise Institute, 87–124.

Heidenheimer, A., Heclo, H. and Teich Adams, C. (1990) *Comparative Public Policy*, New York: St Martin's Press.

Heiner, R. A. (1986) Uncertainty, signal-detection experiments, and modeling behavior, in R. N. Langlois (ed.) *Economics as a Process, Essays in the New Institutional Economics*, Cambridge: Cambridge University Press, 59–115.

Héritier, A. and associates (1996) *Ringing The Changes in Europe: Regulatory Competition and the Transformation of the State, Britain, France, Germany*, Berlin: de Gruyter.

Hicks, D. M. and Katz, S. (1996) Where is science going?, *Science, Technology, and Human Values* 21: 379–406.

Hilgartner, S. (1990) The dominant view of popularization: conceptual problems, political uses, *Social Studies of Science* 20: 519–39.

Hirschman, A. O. (1970) *Exit, Voice, and Loyalty, Responses to Decline in Firms, Organizations, and States*, Cambridge, MA: Harvard University Press.

Hirschman, A. O. (1977) *The Passions and the Interests, Political Arguments for Capitalism before its Triumph*, Princeton, NJ: Princeton University Press.

Hirschman, A. O. (1982) *Shifting Involvements: Private Interest and Public Action*, Princeton, NJ: Princeton University Press.

Hohn, H. W. and Schneider, V. (1991) Path-dependency and critical mass in the development of research and technology: a focused comparison, *Science and Public Policy* 18: 111–22.

Hollingsworth, J. R. (2000) Strategies for doing institutional analysis, in N. Stehr and P. Weingart (eds) *Practising Interdisciplinarity*, Toronto: University of Toronto Press.

Holton, G. (1994) On doing one's damnedest: the evolution of trust in scientific findings, in D. H. Guston and K. Keniston (eds) *The Fragile Contract, University Science and the Federal Government*, Cambridge, MA: The MIT Press, 59–81.

Houghton, John T. *et al.* (eds) (1996) *Climate Change 1995: The Science of Climate Change*. Cambridge: Cambridge University Press.

Huber, M. and Liberatore, A. (2000) A regional approach to the management of global environmental risks. The case of the European Community, in W. Clark, J. Jäger and J. van Eijndhoven (eds) *Learning to Manage Global Environmental Risks: A Comparative History of Social Responses to Climate Change, Ozone Depletion and Acid Rain*. Cambridge, MA: The MIT Press.

Hucke, J. (1990) Umweltpolitik: Die Entwicklung eines neuen Politikfelds, in K. von Beyme and M. G. Schmidt (eds) *Politik in der Bundesrepublik Deutschland*, Opladen: Westdeutscher Verlag, 382–98.

Hughes, E. E., Dorko, W. D. and Taylor, J. K. (1978) *Evaluation of Methodology for Analysis of Halocarbons in the Upper Atmosphere: Phase 1*, NBSIR 78–1480, Washington, DC: National Bureau of Standards, Department of Commerce.

Hunter, J. S. (1980) The national system of scientific measurement, *Science* 210: 869–74.

IMOS (1975) Fluorocarbons and the environment, in Council on Environmental Quality/ Federal Council for Science and Technology (ed.) *Report of Federal Task Force on Inadvertent Modification of the Stratosphere (IMOS)*, Washington, DC: US Government Printing Office.

Irwin, A. (1995) *Citizen Science. A Study Of People, Expertise And Sustainable Development*, London: Routledge.

Irwin, A. and Wynne, B. (eds) (1996) *Misunderstanding Science? The Public Reconstruction of Science and Technology*, Cambridge: Cambridge University Press.

Irwin, A. *et al.* (1998) Regulatory science – towards a sociological framework, *Futures* 29: 17–31.

Iyengar, S. (1987) Television news and citizens' explanations of national affairs, *American Political Science Review* 81: 815–31.

Jachtenfuchs, M. (1990) The European Community and the protection of the ozone layer, *Journal of Common Market Studies* 28: 261–77.

Jachtenfuchs, M. (1995) Ideen und internationale Beziehungen, *Zeitschrift für Internationale Beziehungen* 2: 417–42.

Jasanoff, S. (1986) *Risk Management and Politcal Culture, A Comparative Study of Science in the Policy Context*, New York: Sage.

Jasanoff, S. (1990) *The Fifth Branch, Science Advisers as Policymakers*, Cambridge, MA: Harvard University Press, 208–50.

Jasanoff, S. (1992) Science, politics, and the renegotiation of expertise at EPA, *OSIRIS* 7: 195–217.

Jasanoff, S. (1995) *Science at the Bar: Law, Science, and Technology in America*, Cambridge, MA: Harvard University Press.

Jasanoff, S. and Wynne, B. (1998) Science and decision making, in S. Reyner and E. L. Malone (eds) *Human Choice and Climate Change, Vol. 1: The Societal Framework*, Columbus, OH: Batelle Press: 1–87.

Jasanoff, S. *et al.* (eds) (1995) *Handbook of Science and Technology Studies*, London: Sage.

Jesson, J. P. (1982) Halocarbons, in F. A. Bower and R. B. Ward (eds) *Stratospheric Ozone and Man*, Boca Raton: CRC Press, 30–63.

Johnston, H. (1971) Reduction of stratospheric ozone by nitrogen oxide catalysts from supersonic transport, *Science* 173: 517.

Jönsson, S. A. and Lundin, R. (1977) Myths and wishful thinking as management tools, in P. C. Nystrom and W. H. Starbuck (eds) *Prescriptive Models of Organizations*, Amsterdam: North Holland, 157–70.

Jones, B. D. (1994) *Reconceiving Decision-Making in Democratic Politics. Attention, Choice, and Public Policy*. Chicago: The University of Chicago Press.

Jungermann, H., Kasperson, R. and Wiedemann, P. (1988) *Risk Communication*, Jülich: KFA Jülich.

Kahneman, D., Slovic, P. and Tversky. A. (1982) *Judgement Under Uncertainty, Heuristics and Biases*, Cambridge: Cambridge University Press.

Kaiser, K. (1969) Transnationale Politik, in Ernst-Otto Czempiel (ed.) *Die anachronistische Souveränität*, Opladen: Westdeutscher Verlag, 80–109.

Katz, E. and Lazarsfeld, P. F. (1955) *Personal Influence: The Part Played by People in the Flow of Mass Communications*, New York: The Free Press.

Kaufmann, F. X. (1987) Interdisziplinäre Wissenschaftspraxis. Erfahrungen und Kriterien, in J. Kocka (ed.) *Interdisziplinarität*, Frankfurt am Main: Suhrkamp, 63–81.

Kenis, P. and Schneider, V. (1991) Policy networks and policy analysis: scrutinizing a new analytical toolbox, in B. Marin and R. Mayntz (eds) *Policy Networks, Empirical Evidence and Theoretical Considerations*, Frankfurt am Main/Boulder, CO: Campus/Westview, 25–59.

Keohane, R. O. (1984) *After Hegemony, Cooperation and Discord in International Political Economy*, Princeton, NJ: Princeton University Press.

Keohane, R. O. (1986) Reciprocity in international relations, *International Organization* 40: 1–27.

Keohane, R. O. and Nye, J. S. (1971) Introduction, in R. O. Keohane and J. S. Nye (eds) *Transnational Relations and World Politics*, Cambridge, MA: Harvard University Press, xii–xvi.

Keohane, R. O. and Ostrom, E. (1994) Local commons and global interdependence: heterogeneity and cooperation in two domains, *Journal of Theoretical Politics* 6: 403–28.

Keplinger, H. M. (1988) *Künstliche Horizonte*, München: Oldenbourg.

Kerr, R. A. (1987) Halocarbons linked to ozone hole, *Science* 236: 1182–3.

Kerr, R. A. (1988) Evidence of Arctic ozone destruction, *Science* 240: 1144–5.

Keynes, J. M. (1936) *The General Theory of Employment, Interest and Money*, London: Macmillan.

Kingdon, J. W. (1984) *Agendas, Alternatives, and Public Policies*, Boston, MA: Little Brown.

Klein, D. B. (1997) Knowledge, reputation, and trust, by voluntary means, in D. Klein (ed.) *Reputation. Studies in the Voluntary Eliticiation of Good Conduct*, Ann Arbor: The University of Michigan Press.

Klein, J. T. and Porter, A. L. (1990) Preconditions for interdisciplinary research, in P. Birnbaum-More, F. A. Rossini and D. R. Baldwin (eds) *International Research Management. Studies in Interdisciplinary Methods from Business, Government and Academia*, New York: Oxford University Press.

Knight, F. (1921) *Risk, Uncertainty and Profit*, New York: Houghton Mifflin.

Knorr-Cetina, K. (1981) *The Manufacture of Knowledge, an Essay on the Constructivist and Contextual Nature of Science*, Oxford: Pergamon.

Krasner, S. (1976) State power and the structure of international trade, *World Politics* 38: 317–43.

Krasner, S. (1983a) Structural causes and regime consequences: regimes as intervening variables, in S. Krasner (ed.) *International Regimes*, London: Cornell University Press, 1–21.

Krasner, S. (1983b) Regimes and the limits of realism: regimes as autonomous variables, in S. Krasner (ed.) *International Regimes*, London: Cornell University Press, 355–68.

Krohn, W. and Küppers, G. (1989) *Die Selbstorganisation der Wissenschaft*, Frankfurt am Main: Suhrkamp.

Krueger, J. and Rowlands, I. (1996) Institutions for global environmental change, *Global Environmental Change* 6: 245–7.

Kuhn, T. S. (1970) *The Structure of Scientific Revolutions*, Chicago, IL: Chicago University Press.

Kuhn, T. S. (1977) *The Essential Tension: Selected Studies in Scientific Tradition and Change*, Chicago, IL: University of Chicago Press.

Küppers, G., Lundgreen, P. and Weingart. P. (1978) *Umweltforschung – die gesteuerte Wissenschaft?*, Frankfurt am Main: Suhrkamp.

LaFollette, M. C. (1992) *Stealing into Print, Fraud, Plagiarism, and Misconduct in Scientific Publishing*, Berkeley, CA: University of California Press.

Lakoff, G. and Johnson, M. (1980) *Metaphors we Live By*, Chicago, IL: Chicago University Press.

Lambright, W. H. (1995) NASA, ozone, and policy-relevant science, *Research Policy* 24: 747–60.

Landy, M. K., Roberts, M. J. and Thomas, S. (1990) *The Environmental Protection Agency: Asking the Wrong Questions*, New York: Oxford University Press.

Lang, W. (1988) Diplomatie zwischen Ökonomie und Ökologie, *Europa-Archiv*, Folge 4: 105–10.

Lang, W. (1989) *Internationaler Umweltschutz, Völkerrecht und Außenpolitik zwischen Ökonomie und Ökologie*, Wien: Orac.

Lang, W. (1991) Is the ozone depletion regime a model for an emerging regime on global warming?, *UCLA Journal of Environmental Law* 9: 161–74.

Lang, W. (1994) Environmental treatymaking: lessons to be learned for controlling pollution of outer space, in John Simpson (ed.) *Preservation for Near-Earth Space for Future Generations*, Cambridge: Cambridge University Press, 165–79.

Lash, S., Szerzynsky, B. and Wynne, B. (eds) (1996) *Risk, Environment and Modernity*, London: Sage.

Latour, B. (1983) Give me a laboratory and I will raise the world, in K. Knorr-Cetina and M. Mulkay (eds) *Science Observed: Perspectives on the Social Study of Science*, London: Sage, 141–70.

Latour, B. (1987) *Science in Action, How to Follow Scientists and Engineers Through Society*, Milton Keynes: Open University Press.

Latour, B. (1990) Drawing things together, in M. Lynch and S. Woolgar (eds) *Representation in Scientific Practice*, Cambridge, MA: The MIT Press, 19–68.

Latour, B. (1993) *We Have Never Been Modern*, Cambridge, MA: Harvard University Press.

Latour, B. and Woolgar, S. (1986) *Laboratory Life, The Construction of Scientific Facts*, Princeton, NJ: Princeton University Press.

Latour, B., Maugin, P. and Teil, G. (1992) A note on socio-technical graphs, *Social Studies of Science* 22: 33–57.

Lau, C. (1989a) Die Definition gesellschaftlicher Probleme durch die Sozialwissenschaften, in U. Beck and W. Bonß (eds) *Weder Sozialtechnologie noch Aufklärung? Analysen zur Verwendung sozialwissenschaftlichen Wissens*, Frankfurt am Main: Suhrkamp, 384–419.

Lau, C. (1989b) Risikodiskurse: Gesellschaftliche Auseinandersetzungen um die Definition von Risiken, *Soziale Welt* 40: 418–36.

Lau, R., Smith, R. and Fiske, S. (1991) Political beliefs, policy interpretations, and political persuasion, *Journal of Politics* 53: 644–75.

Lepenies, W. (1989) Die Idee der deutschen Universität – Aus der Sicht der Wissenschaftsforschung, in W. Lepenies *Gefährliche Wahlverwandtschaften, Essays zur Wissenschaftsgeschichte*, Stuttgart: Reclam, 140–60.

Levy, D. L. (1996) Business and international treaties: ozone depletion and climate change, *California Management Review* 39(3): 54–71.

Limbaugh, R. (1992) *The Way Things Ought to Be*, New York: Simon and Schuster.

Litfin, K. T. (1994) *Ozone Discourses: Sciences and Politics in Global Environmental Cooperation*, New York: Columbia University Press.

Lotspeich, R. (1998) Comparative environmental policy, *Policy Studies Journal* 26: 85–104.

Lovelock, J. E. (1974) Atmospheric halocarbons and stratospheric ozone, *Nature* 252: 292–4.

Lovelock, J. E. (1979) *Gaia – A New Look at Life on Earth*, Oxford: Oxford University Press.

Lovelock, J. E. (1982) Epilogue, in F. A. Bower and R. B. Ward (eds) *Stratospheric Ozone and Man*, Boca Raton: CRC Press, 241–53.

Lovelock, J. E. (1984) Causes and effects of changes in stratospheric ozone: update 1983 (review), *Environment* 26(10): 25–6.

Lovelock, J. E. (1988) *The Ages of Gaia, A Biography of Our Living Earth*, New York: W. W. Norton and Co.

Lovelock, J. E., Maggs, R. J. and Wade, R. J. (1973) Halogenated hydrocarbons in and over the Atlantic, *Nature* 241: 194–6.

Lubinska, A. (1985) Europe takes a cheerful view, *Nature* 313: 727.

Luce, D. and Raiffa, H. (1957) *Games and Decisions*, New York: John Wiley & Sons.

Luhmann, H.-J. (1991) Warum hat nicht der Sachverständigenrat für Umweltfragen, sondern der *Spiegel* das Waldsterben entdeckt?, in G. Altner *et al.* (eds) *Jahrbuch Ökologie 1992*, München: Beck, 292–307.

Luhmann, N. (1995) *Social Systems* (translated by J. Bednarz, Jr. with D. Baecker, foreword by Eva M. Knodt), Stanford, CA: Stanford University Press.

Luhmann, N. (1989) *Ecological Communication*, Cambridge: Polity.

Lukes, S. (1974) *Power – A Radical View*, London: Macmillan.

McElroy, M. (1986) Reduction of Antarctic ozone due to synergistic interactions of chlorine and bromide, *Nature* 321: 759–62.

McGarity, T. O. (1991) *Reinventing Rationality, The Role of Regulatory Analysis in the Federal Bureaucracy*, Cambridge: Cambridge University Press.

McInnis, D. F. (1992) Ozone layers and oligopoly profits, in M. Greve and F. Smith (eds) *Environmental Politics*, New York: Praeger, 129–54.

MacKenzie, D. (1993) Negotiating artihmetic, constructing proof: the sociology of mathematics and information technology, *Social Studies of Science* 23: 37–66.

MacNaghten, P. and Urry, J. (1998) *Contested Natures*, London: Sage.

Maduro, R. and Schauerhammer, R. (1992) *The Holes in the Ozone Scare: The Scientific Evidence that the Sky Isn't Falling*, Washington, DC: 21st Century Science Associates.

Majone, G. (1989) *Evidence, Argument and Persuasion in the Policy Process*, New Haven, CT: Yale University Press.

Majone, G. (1993) Wann ist Policy-Deliberation wichtig?, in A. Héritier (ed.) *Policy-Analyse, Kritik und Neuorientierung*, Politische Vierteljahresschrift, Sonderheft 24: 97–115.

Majone, G. (1996a) *Regulating Europe*, London: Routledge.

Majone, G. (1996b) Public policy and administration: ideas, interests and institutions, in R. E. Goodin and H.-D. Klingemann (eds) *A New Handbook of Political Science*, Oxford: Oxford University Press, 610–27.

Malkin, J. and Wildavsky, A. (1991) Why the traditional distinction between private and public goods should be abandoned, *Journal of Theoretical Politics* 3: 355–78.

Manufacturing Chemists Association (1975) *Research Program on Effect of Fluorcarbons on the Atmosphere*, Washington, DC: MCA.

March, J. G. and Simon, H. A. (1958) *Organizations*, New York: John Wiley & Sons.

Marcus, A. A. (1980) Environmental Protection Agency, in J. Q. Wilson (ed.) *The Politics of Regulation*, New York: Basic Books, 267–303.

Marcus, A. A. (1988) Risk, uncertainty, and scientific judgement, *Minerva* 26: 138–52.

Marcus, A. A. (1991) EPA's organizational structure, *Law and Contemporary Problems* 54: 5–40.

Martell, L. (1994) *Ecology and Society. An Introduction*, Cambridge: Polity.

Maruyama, M. (1963) The second cybernetics: deviation amplifying mutual causal processes, *Scientific American* 51: 164–79.

Maxwell, J. H. and Weiner, S. L. (1993) Green consciousness or dollar diplomacy? The British response to the threat of ozone depletion, *International Environmental Affairs* 5: 19–41.

Mayntz, R. (1993) Policy-Netzwerke und die Logik von Verhandlungssystemen, in A. Héritier (ed.) *Policy-Analyse, Kritik und Neuorientierung*, Politische Vierteljahresschrift, Sonderheft 24: 39–56.

Mayntz, R. and Nedelmann, B. (1987) Eigendynamische soziale Prozesse. Anmerkungen zu einem analytischen Paradigma, *Kölner Zeitschrift für Soziologie und Sozialpsychologie* 39: 648–68.

Mayntz, R. and Scharpf, F. W. (eds) (1995) *Gesellschaftliche Selbstregelung und politische Steuerung*, Frankfurt am Main: Campus.

Mayntz, R. *et al.* (1978) *Vollzugsprobleme der Umweltpolitik*, Materialien zur Umweltforschung, Wiesbaden.

Mazur, A. (1981) *The Dynamics of Technical Controversy*, Washington, DC: Communications Press.

Mazur, A. (1989) Allegations of dishonesty in research and their treatment by American universities, *Minerva* 27: 177–94.

Mazur, A. and Lee, J. (1993) Sounding the global alarm: environmental issues in the US national news, *Social Studies of Science* 23: 681–720.

Meadows, D. and Meadows, D. (1972) *The Limits to Growth*, London and Sydney: Pan Books.

Menke-Glückert, P. (1990) Ökologische Eckwerte als Instrument der Umweltplanung, in W. Schenkel and P. Christoph Storm (eds) *Umwelt: Politik, Technik, Recht, Heinrich von Lersner zum 60, Geburtstag*, Berlin: Erich Schmidt.

Merton, R. K. (1936) The unanticipated consequences of purposive social action, *American Journal of Sociology* 1: 894–904.

Merton, R. K. (1957) *Social Theory and Social Structure*, 2nd edn, Glencoe: Free Press.

Merton, R. K. (1973a) [1942]: The normative structure of science, in R. K. Merton *The Sociology of Science, Theoretical and Empirical Investigations*, Chicago, IL: University of Chicago Press, 267–78.

Merton, R. K. (1973b) [1963]: The ambivalence of scientists, in R. K. Merton *The Sociology of Science, Theoretical and Empirical Investigations*, Chicago, IL: University of Chicago Press, 383–412.

Merton, R. K. (1973c) [1968]: The Matthew Effect in science, in R. K. Merton *The Sociology of Science, Theoretical and Empirical Investigations*, Chicago, IL: University of Chicago Press, 439–59.

Merton, R. K. (1995) The Thomas Theorem and the Matthew Effect, *Social Forces* 74: 379–424.

Milbrath, L. (1984) *Environmentalists, Vanguard for a New Society*, Buffalo: State University of New York Press.

Mills, C. W. (1961) *The Sociological Imagination*, New York: Grove Press.

Mills, C. W. (1963) *Power, Politics and People*, New York: Ballantine Books.

Mitchell, R. C., Mertig, A. and Dunlap, R. (1992) *Twenty Years of Environmental Mobilization: Trends Among National Environmental Organizations*, Washington, DC: Taylor & Francis, 11–26.

Mitroff, I. (1974) *The Subjective Side of Science*, Amsterdam: Elsevier.

Moe, T. M. (1989) The politics of bureaucratic structure, in J. E. Chubb and P. E. Peterson (eds) *Can the Government Govern?* Washington, DC: The Brookings Institution, 267–329.

Moe, T. M. and Caldwell, M. (1994) The institutional foundations of democratic government: a comparison of presidential and parliamentary systems, *Journal of Institutional and Theoretical Economics* 150(1): 171–95.

Molina, M. J. and Rowland, F. S. (1974) Stratospheric sink for chlorofluormethanes: chlorine-atom catalysed destruction of ozone, *Nature* 249 (28 June): 810–12.

Molina, M. J. and Molina, L. (1986) Production of Cl_2O_2 by the self reaction of the ClO radical, *Journal of Physical Chemistry* 91: 433.

Montzka, S. A. *et al.* (1996) Decline in the trospospheric abundance of halogen from halocarbons: implications for stratospheric ozone depletion, *Science* 272: 1318–22.

Morrisette, P. M. (1989) The evolution of policy responses to stratospheric ozone depletion, *Natural Resources Journal* 29: 793–820.

Moss, R. (1995) The IPCC: policy relevant (not driven) scientific assessment. A comment on Sonja Boehmer-Christiansen's: 'Global climate protection policy: the limits of scientific advice', *Global Environmental Change* 5: 171–4.

Mukerji, C. (1989) *A Fragile Power, Scientists and the State*, Princeton, NJ: Princeton University Press.

Mulkay, M. (1969) Some aspects of cultural growth in the natural sciences, *Social Research* 36: 22–52.

Müller, E. (1986) *Innenwelt der Umweltpolitik*, Opladen: Westdeutscher Verlag.

Müller, E. (1989) Sozial-liberale Umweltpolitik. Von der Karriere eines neuen Politikbereichs, *Aus Politik und Zeitgeschichte, Beilage zum Parlament* B47–48/89: 3–15.

Murphy, R. (1997) *Sociology and Nature. Social Action in Context*, Boulder, CO: Westview.

Musgrave, R. A. and Musgrave, P. B. (1989) *Public Finance in Theory and Practice*, 5th edn, New York and London: McGraw-Hill.

Nance, J. (1991) *What Goes Up, The Global Assault on Our Atmosphere*, New York: William Morrow.

NASA (National Aeronautics and Space Administration) (1977) Effects of chlorofluoromethanes on stratospheric ozone. Assessment report. September, NASA: Ames Research Center.

NASA (1987) *Airborne Antarctic Ozone Experiment*, NASA: Ames Research Center.

Nelkin, D. (ed.) (1979) *Controversy: Politics of Technical Decisions*, Beverly Hills, CA: Sage.

Newman, P. (1994) Antarctic total ozone in 1958, *Science* 264: 543–6.

Norgaard, Richard B. (1995) *Development betrayed. The end of progress and a coevolutionary revisioning of the future*. London: Routledge.

Oberthür, S. (1999) The EU as an international actor: the protection of the ozone layer, *Journal of Common Market Studies* 37(4): 641–59.

O'Connell, J. (1993) Metology: the creation of universality by the circulation of particulars, *Social Studies of Science* 23: 129–73.

Olson, M. Jr. (1965) *The Logic of Collective Action*, Cambridge, MA: Harvard University Press.

Ophuls, W. (1973) Leviathan or oblivion, in H. E. Daly (ed.) *Toward a Steady State Economy*, San Francisco, CA: W. H. Freeman.

O'Riordan, T. (1971) *Environmentalism*, London: Pion.

O'Riordan, T. and Jordan, A. (1999) Institutions, climate change and cultural theory, *Global Environmental Change* 9: 81–93.

O'Riordan, T. and Wynne, B. (1987) Regulating environmental risks: a comparative perspective, in P. R. Kleindorfer and H. C. Kunreuther (eds) *Insuring and Managing Hazardous Risks*, Berlin: Springer, 389–410.

O'Riordan, T., Cooper, C. L., Jordan, A., Rayner, S., Richards, K. R., Runci, P. and Yoffe, S. (1998) Institutional frameworks for political action, in S. Reyner and E. L. Malone (eds) *Human Choice and Climate Change, Vol. 1: The Societal Framework*, Columbus, OH: Batelle Press, 345–439.

Ostrom, E. (1990) *Governing the Commons: The Evolution of Institutions for Collective Action*, New York: Cambridge University Press.

Ostrom, E. (1994) Constituting social capital and collective action, *Journal of Theoretical Politics* 6: 527–62.

Ostrom, E., Gardner, R. and Walker, J. (1994) *Rules, Games and Common-pool Resources*, Ann Arbor: University of Michigan Press.

Otway, H. (1985) Introduction, in H. Otway and M. Peltu (eds) *Regulating Industrial Risks: Science, Hazards and Public Protection*, London: Butterworth.

Otway, H. and Thomas, K. (1982) Reflections on risk perception and policy, *Risk Analysis* 2: 69–82.

Oye, K. A. (1986) Explaining cooperation under anarchy: hypotheses and strategies, in K. A. Oye (ed.) *Cooperation Under Anarchy*, Princeton, NJ: Princeton University Press, 1–24.

Oye, K. A. and Maxwell, J. (1994) Self-interest and environmental management, *Journal of Theoretical Politics* 6: 593–624.

Parson, E. A. (1993) Protecting the ozone layer, in Peter Haas *et al.* (eds) *Institutions for the Earth, Sources of Effective International Environmental Protection*, Cambridge, MA: The MIT Press, 27–73.

Parson, E. A. and Greene, O. (1995) The complex chemistry of the international ozone agreements, *Environment* 37: 16–43.

Peters, B. (1996) *Prominenz, Eine soziologische Untersuchung ihrer Entstehung und Wirkung*, Opladen: Westdeutscher Verlag.

Peters, H. P. (1994) Wissenschaftliche Experten in der öffentlichen Kommunikation über Technik, Umwelt und Risiken, in F. Neidhardt (ed.) *Öffentlichkeit, öffentliche Meinung, soziale Bewegungen*, Kölner Zeitschrift für Soziologie und Sozialpsychologie, Sonderheft 34: 162–90.

Pielke, R. Jr. and Betsill, M. (1997) Policy for science for policy: a commentary on Lambright on ozone depletion and acid rain, *Research Policy* 26: 157–68.

Pizzorno, A. (1986) Some other kind of otherness: a critique of 'rational choice' theories, in A. Foxley and M. S. McPherson (eds) *Development, Democracy, and the Art of Trespassing*, Notre Dame, IN: University of Notre Dame Press, 355–73.

Polanyi, M. (1958) *Personal Knowledge*, London: Routledge and Kegan Paul.

Porter, A. L. and Rossini, F. A. (1985) Peer review of interdisciplinary research proposals, *Science, Technology and Human Values* 10: 33–8.

Porter, T. M. (1995) *Trust in Numbers, The Pursuit of Objectivity in Science and Public Life*, Princeton, NJ: Princeton University Press.

Powell, W. W. (1990) Neither market nor hierarchy: network forms of organization, *Research in Organizational Behaviour* 12: 295–336.

Pratt, J. W. and Zeckhauser, R. J. (eds) (1985) *Principals and Agents: The Structure of Business*, Boston, MA: Harvard Business School Press.

Press, E. and Washburn, J. (2000) The kept university, *Atlantic Monthly* (March). http://www.theatlantic.com/issues/2000/03/press.htm

Price, D. (1963) *Little Science, Big Science . . . and Beyond*, New York: Columbia University Press.

Price, D. (1970) Citation measures of hard science, soft science, technology, and non-science, in C. E. Nelson and D. K. Pollack (eds) *Communication Among Scientists and Engineers*, Lexington, MA: Heath, 3–22.

Pruit, D. G. and Lewis, S. (1977) The psychology of integrative bargaining, in D. Druckman (ed.) *Negotiations*, Beverly Hills, CA: Sage, 161–92.

Przeworski, A. and Teune, H. (1970) *The Logic of Comparative Social Inquiry*, New York: Wiley-Interscience.

Pukelsheim, F. (1990) Robustness of statistical gossip and the Antarctic ozone hole (Letter to the Editor), *IMS Bulletin* 19: 540–5.

Purvis, M., *et al.* (1997) Fragmenting uncertainties: some British business responses to stratospheric ozone depletion, *Global Environmental Change* 7(2): 93–111.

Putnam, R. D. (1988) Diplomacy and domestic politics: the logic of two-level games, *International Organization* 42: 427–60.

Rat von Sachverständigen für Umweltfragen (1978) *Umweltgutachten 1978*, Stuttgart: W. Kohlhammer.

Ravetz, J. (1987) Uncertainty, ignorance and policy, in H. Brooks and C. L. Cooper (eds) *Science for Public Policy*, Oxford: Pergamon.

Rawls, J. (1971) *A Theory of Justice*, Oxford: Oxford University Press.

Ray, D. L. and Guzzo, L. (1990) *Trashing the Planet*, Washington, DC: Regnery Gateway.

Redclift, M. and T. Benton (1994) *Social Theory and Global Environment*. London: Routledge.

Redclift, M. and G. Woodgate (1994) Sociology and the Environment: Discordant Discourse? in M. Redclift and T. Benton (eds) *Social Theory and the Global Environment*. London: Routledge, 51–66.

Rein, M. and Schön, D. (1993) Reframing policy discourse, in F. Fischer and J. Forester (eds) *The Argumentative Turn in Policy Analysis and Planning*, Durham, NC: Duke University Press, 145–66.

Reinhardt, F. (1989) *Du Pont Freon Products Division (A)*, Harvard Business School Case Study, Washington, DC: National Wildlife Federation.

Renn, O. (1992) The social arena concept of risk debates, in S. Krimsky and D. Golding (eds) *Social Theories of Risk*, Westport, CT: Praeger, 179–96.

Renn, O. (1996) Rolle und Stellenwert der Soziologie in der Umweltforschung, in A. Diekmann and C. Jäger (eds) *Umweltsoziologie*, Kölner Zeitschrift für Soziologie und Sozialpsychologie, Sonderheft 36: 28–58.

Restivo, S. (1988) Science as a social problem, *Social Problems* 35(3): 206–25.

Riker, W. H. (1964) *The Theory of Political Coalitions*, New Haven, CT: Yale University Press.

Riker, W. H. (1984) The heresthetics of constitution-making: the presidency in 1787, with comments on determinism and rational choice, *American Political Science Review* 78: 1–16.

Riker, W. H. (1986) *The Art of Political Manipulation*, New Haven, CT: Yale University Press.

Risse-Kappen, T. (1994) Ideas do not float freely: transnational coalitions, domestic structures, and the end of the cold war, *International Organization* 48: 185–214.

Risse-Kappen, T. (1995) Bringing transnational relations back in: introduction, in T. Risse-Kappen, *Bringing Transnational Relations Back in Non-State-Actors, Domestic Structures, and International Institutions*, New York: Cambridge University Press, 3–33.

Rittberger, V. (ed.) (1993) *Regime Theory and International Relations*, Oxford: Clarendon.

Roan, S. (1989) *Ozone crisis, The 15-Year Evolution of a Sudden Global Emergency*, New York: Wiley.

Rogers, E. M. (1962) *Diffusion of Innovations*, New York: The Free Press.

Rosenau, J. N. and Czempiel, E. O. (eds) (1992) *Governance Without Government: Order and Change in World Politics*, New York: Cambridge University Press.

Rosenbaum, W. A. (1994) The clenched fist and the open hand: into the 1990s at EPA, in N. J. Vig and M. E. Kraft (eds) *Environmental Policy in the 1990s*, Washington, DC: CQ Press, 121–43.

Rowland, F. S. (1993) President's Lecture: The need for scientific communication with the public, *Science* 260: 1571– 6.

Rowlands, I. H. (1995) *The Politics of Global Atmospheric Change*, Manchester: Manchester University Press.

Russow, J. (1977) Die FKW-Ozon-Hypothese, *Nachrichten aus Chemie und Technik* 25: 507ff.

Sabatier, P. A. (1993) Policy change over a decade or more, in P. A. Sabatier and H. C. Jenkins-Smith (eds) *Policy Change and Learning, An Advocacy Coalition Approach*, Boulder, CO: Westview Press, 13–40.

Sabatier, P. A. and Jenkins-Smith, H. C. (eds) (1993) *Policy Change and Learning, An Advocacy Coalition Approach*, Boulder, CO: Westview Press.

Sachs, W. (1999) *Planet Dialectics*, London: Zed.

Sachs, W., Loske, R. and Linz, M. (1998) *Greening the North, A Post-Industrial Blueprint for Ecology and Equity*, London: Zed.

Salter, J. R. (ed.) (1996) *European Environmental Law*, Vol. 1, London: Kluwer.

Sand, P. H. (ed.) (1992) *The Effectiveness of International Environmental Agreements: A Survey of Existing Legal Instruments*, Cambridge: Grotius Publications.

Sbragia, A. M. (1999) The changing role of the European Union in international environmental politics: institution building and the politics of climate change, *Environment and Planning Government and Policy* 17: 53–68.

Schaffer, S. (1998) The Leviathan of Parsontown: literary technology and scientific representation, in T. Lenoir (ed.) *Inscribing Science*, Stanford, CA: Stanford University Press, 182–222.

Scharpf, F. W. (1988) The joint decision trap, *Public Administration* 66: 239–87.

Scharpf, F. W. (1989) Politische Steuerung und politische Institutionen, *Politische Vierteljahresschrift* 30: 10–21.

Scharpf, F. W. (1993) Coordination in hierarchies and networks, in F. W. Scharpf (ed.) *Games in Hierarchies and Networks. Analytical and Empirical Approaches to the Study of Governance Institutions*, Frankfurt am Main/New York: Campus/Westview, 125–65.

Scharpf, F. W. and Mohr, M. (1994) *Efficient Self-Coordination in Policy Networks, A Simulation Study*, MPIfG Discussion Paper 94/1. Köln: Max-Planck-Institut für Gesellschaftsforschung.

Schattschneider, E. E. (1960) *The Semi-Sovereign People*, Hinsdale: Dryden.

Schelling, T. C. (1960) *The Strategy of Conflict*, New York: Oxford University Press.

Schmidt, S. and Werle, R. (1998) *Coordinating Technology: Studies in the International Standardization of Telecommunications*, Cambridge, MA: The MIT Press.

Schnaiberg, A., Watts, N. and Zimmermann, K. (eds) (1986) *Distributional Conflicts in Environmental Policy*, Aldershot: Gower.

Schneider, V. (1988) *Politiknetzwerke der Chemikalienkontrolle, Eine Analyse einer transnationalen Politikentwicklung*, Berlin: de Gruyter.

Schoeberl, M. R. (1988) Dynamics weaken the polar hole, *Nature* 336: 420–1.

Schon, D. A. (1982) The fear of innovation, in S. B. Barnes and D. O. Edge (eds) *Science in Context*, London: Open University Press, 290–302.

Schubert, A. and Braun, T. (1990) World flash on basic research. International collaboration in the sciences (1981–1985), *Sociometrics* 19: 3–10.

Schulz, W. (1976) *Die Konstruktion von Realität in den Nachrichtenmedien, Analyse der aktuellen Berichterstattung*, Freiburg: Alber.

Scott, A. (1990) *Ideology and the New Social Movements*, London: Unwin Hyman.

Sebenius, J. K. (1992) Challenging conventional explanations of international cooperation:

negotiation analysis and the case of epistemic communities, *International Organization* 46: 323–65.

Sen, A. (1992) On the Darwinian view of progress, *London Review of Books*, 5 November: 15–19.

Shackley, S. (1997) The intergovernmental panel on climate change: consensual knowledge and global politics, *Global Environmental Change* 7: 77–9.

Shackley, S. and Skodvin, T. (1995) IPCC gazing and the interpretative social sciences. A comment on Sonja Boehmer-Christiansen's: 'Global climate protection policy: the limits of scientific advice', *Global Environmental Change* 5: 175–80.

Shapin, S. (1984) Pump and circumstance: Robert Boyle's literary technology, *Social Studies of Science* 14: 481–520.

Shapin, S. (1994) *A Social History of Truth, Civility and Science in Seventeenth-Century England*, Chicago, IL: University of Chicago Press.

Shell, E. (1987) Weather versus chemicals, *Atlantic Monthly* 259: 27–31.

Shepherd, G. B. (ed.) (1995) *Rejected, Leading Economists Ponder the Publication Process*, Sun Lakes, AZ: Thomas Horton and Daughters.

Shils, E. (1987) Science and scientists in the public arena, *The American Scholar* 35: 185–202.

Shiva, V. (1989) *Staying Alive: Women, Ecology, and Development*, London: Zed.

Simon, J. L. and H. Kahn (eds) (1984) *The Resourceful Earth, a Response to Global 2000*. Oxford: Blackwell.

Singer, F. S. (1989) My adventures in the ozone layer, *National Review*: 34–8.

Smith, B. (1998) Ethics of Du Pont's CFC strategy 1975–1995, *Journal of Business Ethics* 17: 557–68.

Smith, V. K. (ed.) (1984) *Environmental Policy under Reagan's Executive Order*, Chapel Hill, NC: The University of North Carolina Press.

Snidal, D. (1979) Public goods, property rights, and political organisations, *International Studies Quarterly* 23(4): 532–66.

Snidal, D. (1986) The game theory of international politics, in K. A. Oye (ed.) *Cooperation Under Anarchy*, Princeton, NJ: Princeton University Press, 25–57.

Snidal, D. (1994) The politics of scope: endogenous actors, heterogeneity and institutions, *Journal of Theoretical Politics* 6: 449–72.

Snow, D. and Benford, R. (1988) Ideology, frame resonance and participant mobilisation, *International Social Movement Research* 1: 197–217.

Solomon, S. (1987) More news from Antarctica, *Nature* 326: 20.

Solomon, S. (1988) The mystery of the Antarctic ozone hole, *Review of Geophysics* 26: 131–48.

Solomon, S., Garcia, R. R., Rowlands, F. S. and Weubbles, D. J. (1986) On the depletion of the Antarctic ozone, *Nature* 321 (19 June): 755– 8.

Soroos, M. (1997) *The Endangered Atmosphere. Preserving a Global Commons*, Columbia, SC: University of South Carolina Press.

Spector, M. and Kitsuse, J. I. (1973) Social problems: a reformulation, *Social Problems* 20: 145–59.

Sprinz, D. and Vaahtoranta, T. (1994) The interest-based explanation of international environmental policy, *International Organization* 48: 77–105.

St. John, D., Bailey, W., Fellner, W., Minor, J. and Sull, R. (1982) Time series analysis of stratospheric ozone, *Communication in Statistics: Theory and Methods* 11: 1293–333.

Starr, C. (1969) Social benefit versus technological risk. What is our society willing to pay for safety?, *Science* 165: 1232–8.

Stehr, N. (1992) *Practical Knowledge*, London: Sage.

Stephan, P. E. (1996) The economics of science, *Journal of Economic Literature* 34: 1199–235.

Stephan, P. E. (1997) Examining the link between science and economic growth, *Scientist* 11: 8.

Stewart, T. R. (1987) The Delphi Technique and judgmental forecasting, in K. C. Land and S. H. Schneider (eds) *Forecasting in the Social and Natural Sciences*, Dordrecht: Reidel, 97–113.

Stigler, G. J. (1971) The theory of economic regulation, *Bell Journal of Economics and Management Science* 2: 1–21.

Stinchcombe, A. L. (1984) The origins of sociology as a discipline, *Acta Sociologica* 27: 51–61.

Stolarski, R. S. and Cicerone, R. (1974) Stratospheric chlorine: a possible sink for ozone, *Canadian Journal of Chemistry* 52: 1610–15.

Stolarski, R. S., Krueger, A. J., Schoeberl, M. R., McPeters, R. D., Newman, P. A. and Alpert, J. C. (1986) Nimbus 7 satellite measurements of the springtime Antarctic ozone decrease, *Nature* 322: 808–11.

Streeck, W. and Schmitter, P. (eds) (1985) *Private Interest Government: Beyond Market and State*, Beverly Hills, CA: Sage.

Tarrow, S. (1992) Mentalities, political culture, and collective action frames: constructing meanings through action, in A. Morris and C. McClurg Mueller (eds) *Frontiers in Social Movement Theory*, New Haven, CT: Yale University Press, 174–202.

Taschner, K. (1987) *The Sky is the Limit, Report of the Seminar on Ozone Depletion and Climate Warming Due to CFCs: The Role of the European Communities*, Brussels: European Environmental Bureau.

Taubes, G. (1993) The ozone backlash, *Science* 260: 1580–3.

Taylor, M. (1987) *The Possibility of Cooperation*, Cambridge: Cambridge University Press.

Temple Lang, J. (1986) The ozone layer convention: a new solution to the question of community participation in 'mixed' international agreements, *Common Market Law Review*: 157–76.

Teubner, G. (1993) The many-headed hydra: networks as higher-order collective actors, in J. McCahery, S. Picciotto and C. Scott (eds) *Corporate Control and Accountability*, Oxford: Clarendon, 41–60.

Thacher, P. (1992) The role of the United Nations, in A. Hurrell and B. Kingsbury (eds) *The International Politics of the Environment*, Oxford: Oxford University Press, 183–211.

Thompson, M. (1983) A cultural basis for comparison, in H. Kunreuther and J. Linneroth (eds) *Risk Analysis and Decision Process*, Berlin: Springer, 233–60.

Thompson, M., Ellis, R. and Wildavsky, A. (1990) *Cultural Theory*, Boulder, CO: Westview.

Timm, G. I. (1989) *Die wissenschaftliche Beratung der Umweltpolitik*, Wiesbaden: Deutscher Universitätsverlag.

Tolba, M. K. (1989) Engagement und Sensibilität, in Mostafa K. Tolba *et al.* (eds) *Die Umwelt bewahren*, Bonn-Bad Godesberg: Stiftung Entwicklung und Frieden, 11–19.

Tolba, M. K. with Rummel-Bulska, Iwona (1998) *Global Environmental Diplomacy*, Cambridge, MA: The MIT Press.

Traweek, S. (1988) *Beamtimes and Lifetimes: The World of High Energy Physicists*, Cambridge, MA: Harvard University Press.

Tsebelis, G. (1990) *Nested Games*, Berkeley, CA: University of California Press.

Umweltbundesamt (1989) *Responsibility Means Doing Without*, Berlin: German Federal Environmental Agency.

Ungar, S. (1992) The rise and (relative) decline of global warming as a social problem, *Sociological Quarterly* 33: 483–501.

Ungar, S. (1998) Bringing the issue back in: comparing the marketability of the ozone hole and global warming, *Social Problems* 45 (4): 510–27.

UKDoE (United Kingdom Department of the Environment) (1976) *Chlorofluorocarbons and their Effect on Stratospheric Ozone*, London: HMSO.

UKDoE (1979) *Chlorofluorocarbons and their Effect on Stratospheric Ozone*, London: HMSO.

UNEP (United Nations Environment Programme) (1987) Ad hoc scientific meeting to compare model generated assessments of ozone layer change for various strategies for CFC control, Manuscript, Nairobi: UNEP.

UNEP (1993) *Register of International Treaties and Other Agreements in the Field of the Environment*, Nairobi: UNEP.

UNEP (1994) *Environmental Effects of Ozone Depletion: 1994 Assessment*, Nairobi: UNEP.

UNEP (1995) *1994 Report of the Technology and Economics Assessment Panel*, Nairobi: UNEP.

US National Academy of Sciences (1976) *Halocarbons: Environmental Effects of Chlorofluoromethane Release*, Washington, DC: National Academy of Sciences.

United States Senate. 94th Congress (1975) *Stratospheric Ozone Depletion, Hearings before the Subcommittee on the Upper Atmosphere of the Committee on Aeronautical and Space Sciences*, Washington, DC: US Government Printing Office.

van den Daele, W., Krohn, W. and Weingart, P. (1979) *Geplante Forschung, Vergleichende Studien über den Einfluß politischer Programme auf die Wissenschaftsentwicklung*, Frankfurt am Main: Suhrkamp.

Vig, N. J, and Kraft, M. E. (1984) Environmental policy from the seventies to the eighties, in N. J. Vig and M. E. Kraft (eds) *Environmental Policy in the 1980s*, Washington, DC: CQ Press, 3–26.

Vig, N. J. and Kraft, M. E. (1994) Environmental policy from the 1970s to the 1990s: continuity and change, in N. J. Vig and M. E. Kraft (eds) *Environmental Policy in the 1990s*, Washington, DC: CQ Press, 3–29.

Viscusi, W. K. (1997) Alarmist decisions with divergent risk information, *Economic Journal* 107: 1657–70.

Voelzkow, H. (1996) *Private Regierungen in der Techniksteuerung*, Frankfurt am Main: Campus.

Vogel, D. (1986) *National Styles of Regulation, Environmental Policy in Great Britain and the United States*, Ithaca: Cornell University Press.

Vogel, D. (1993) Representing diffuse interests in environmental policymaking, in R. K. Weaver and B. A. Rockman (eds) *Do Institutions Matter? Government Capabilities in the United States and Abroad*, Washington, DC: The Brookings Institution, 237–71.

Wagner, R. (1983) The theory of games and the problem of international cooperation, *American Political Science Review* 70: 330–46.

Walker, G. (2000) The hole story, *New Scientist*, 25 March: 24–8.

Walton, R. E. and McKersie, R. B. (1965) *A Behavioral Theory of Labor Negotiations: An Analysis of a Social Interaction System*, New York: McGraw-Hill.

Wasser, H. (1995) Die Interessengruppen, in W. Jäger and W. Welz (eds) *Regierungssystem der USA, Lehr- und Handbuch*, München: Oldenbourg, 297–314.

Weale, A., Pridham, G., Williams, A. and Porter, M. (1996) Environmental administration in six European states: secular convergence or national distinctiveness?, *Public Administration* 74: 255–74.

Weaver, R. K. (1986) The politics of blame avoidance, *Journal of Public Policy* 6: 371–98.

Weidner, H. (1989) Die Umweltpolitik der konservativ-liberalen Regierung, *Aus Politik und Zeitgeschichte, Beilage zum Parlament* B47–48/89: 16–28.

Weinberg, A. (1972) Science and trans-science, *Minerva* 10: 209–22.

Weinberg, A. (1985) Science and its limits: the regulator's dilemma, *Issues in Science and Technology* II, Fall: 68.

Weingart, P. (1974) Das Dilemma: Die Organisation von Interdisziplinarität, *Wirtschaft und Wissenschaft* 3: 22–8.

Weingart, P. (1981) Wissenschaft im Konflikt zur Gesellschaft – Zur De-Institution-alisierung der Wissenschaft, in J. von Kruedener and K. von Schubert (eds) *Technikfolgen und sozialer Wandel, Zur politischen Steuerbarkeit der Technik,* Köln: Verlag Wissenschaft und Politik, 205–24.

Weingart, P. (1983) Verwissenschaftlichung der Gesellschaft – Politisierung der Wissen-schaft, *Zeitschrift für Soziologie* 12: 225–41.

Weingart, P. (1987) Interdisziplinarität als List der Institution, in J. Kocka (ed.) *Interdiszipli-narität,* Frankfurt am Main: Suhrkamp, 159–66.

Weingart, P. (1990) Doomed to passivity? The global ecological crisis and the social sci-ences, in H. Krupp (ed.) *Technikpolitik angesichts der Umweltkatastrophe,* Heidelberg: Physica-Verlag, 48–59.

Weingart, P. and Pansegrau, P. (1999) Reputation in science and prominence in the media: the Goldhagen debate, *Public Understanding of Science* 8: 1–16.

Weingart, P. and Stehr, N. (2000) Introduction, in P. Weingart and N. Stehr (eds) *Practising Interdisciplinarity,* Toronto: University of Toronto Press.

Wettestad, J. and Andresen, S. (1991) *The Effectiveness of International Resource Cooperation: Some Preliminary Findings,* Lysaker: The Fridjof Nansen Institute.

Weyer, J. (1993) System und Akteur: Zum Nutzen zweier soziologischer Paradigmen bei der Erklärung erfolgreichen Scheiterns, *Kölner Zeitschrift für Soziologie und Sozialpsychologie* 45: 1–22.

Wildavsky, A. (1994) *But is it True? A Citizen's Guide to Environmental Health and Safety Issues,* Cambridge, MA: Harvard University Press.

Wilenius, M. (1996) From science to politics: the menace of global environmental change, *Acta Sociologica* 39: 5–30.

Wilkins, L. and Patterson, P. (1990) Risky business: covering slow-onset hazards as rapidly developing news, *Political Communication and Persuasion* 7: 11–23.

Williamson, O. E. (1985) *The Economic Institutions of Capitalism,* New York: The Free Press.

Willke, H. (1983) *Entzauberung des Staates, Überlegungen zu einer sozietalen Gesellschaftstheorie,* Königstein: Athenäum.

Willke, H. (1989) Gesellschaftssteuerung oder partikulare Handlungsstrategien? Der Staat als korporativer Akteur, in M. Glagow *et al.* (eds) *Gesellschaftliche Steuerungsrationalität und partikulare Handlungsstrategien,* Pfaffenweiler: Centaurus, 9–29.

Willke, H. (1995) The proactive state. The role of national enabling policies in global socio-economic transformations, in H. Willke (ed.) *Benevolent Conspiracies,* Berlin: de Gruyter, 325–55.

Wilson, J. Q. (1980) The politics of regulation, in J. Q. Wilson (ed.) *The Politics of Regulation,* New York: Basic Books, 357–94.

Winterhager, M., Weingart, P. and Sehringer, R. (1988) Die Cozitationsanalyse als biblio-metrisches Verfahren zur Messung der nationalen und institutionellen Forschungsper-formanz, in H. -D. Daniel and R. Fisch (eds) *Evaluation von Forschung: Methoden, Ergebnisse, Stellungnahmen,* Konstanz: Universitätsverlag, 319–58.

Wirth, G. F., Brunner, P. W. and Bishop, F. S. (1982) Regulatory actions, in F. A. Bower and R. B. Ward (eds) *Stratospheric Ozone and Man,* Boca Raton: CRC Press, 217–40.

Wolf, C. (1970) The present value of the past, *Journal of Political Economy* 78: 783–92.

Wood, A. (1993) The multilateral fund for the implementation of the Montreal Protocol, *International Environmental Affairs* 5: 335–54.

Woolgar, S. (1981) Interest and explanation in the social study of science, *Social Studies of Science* 11: 365–94.

WMO (World Meteorological Organization) (1986) *Atmospheric Ozone, Assessment of Our Understanding of the Processes Controlling Its Present Distribution and Change*, 3 Bde. Global Ozone Research and Monitoring Project, Report No. 16, Geneva: World Meteorological Organization.

WMO (1988) *Report of the International Ozone Trends Panel*, 2 Bde. Global Ozone Research and Monitoring Project, Report No. 18, Geneva: World Meteorological Organization.

WMO (1989) *Scientific Assessment of Stratospheric Ozone: 1989*, Global Ozone Research and Monitoring Project, Report No. 20, Geneva: World Meteorological Organization.

WMO (1991) *Scientific Assessment of Ozone Depletion: 1991*, Global Ozone Research and Monitoring Project, Report No. 25, Geneva: World Meteorological Organization.

WMO (1993) *Global Atmosphere Watch*, Geneva: World Meteorological Organization.

WMO (1994) *Scientific Assessment of Ozone Depletion: 1994*, Global Ozone Research and Monitoring Project, Report No. 37, Geneva: World Meteorological Organization.

Wuebbles, D. J. (1981) *The Relative Efficiency of a Number of Halocarbons for Destroying Stratospheric Ozone*, Livermore, CA: Lawrence Livermore Laboratory.

Wuebbles, D. J. and Rosenberg, N. (1998) The natural science of global climate change, in S. Rayner and E. L. Malone (eds) *Human Choice and Climate Change*, Vol. 2, Columbus, OH: Batelle Press, 1–78.

Wynne, B. (1992) Uncertainty and environmental learning. Reconceiving the science and policy in the preventive paradigm, *Global Environmental Change*, June: 111–27.

Wynne, B. (1996a) SSK's identity parade: signing-up, off-and-on, *Social Studies of Science* 26: 357–91.

Wynne, B. (1996b) Patronising Joe Public, *The Times Higher Education Supplement*, 12 April.

Yearley, S. (1996) *Sociology, Environmentalism, Globalization: Reinventing the Globe*, London: Sage.

Yin, R. K. (1994) *Case Study Research: Design and Methods*, London: Sage.

Young, O. and Osherenko, G. (eds) (1993) *Polar Politics, Creating International Environmental Regimes*, Ithaca: Cornell University Press.

Young, O. R. (1989) *International Cooperation: Building Regimes for Natural Resources and the Environment*, London: Cornell University Press.

Young, O. R. (1991) Political leadership and regime formation: on the development of institutions in international society, *International Organization* 45: 281–308.

Young, O. R. (1994) *International Governance: Protecting the Environment in a Stateless Society*, Ithaca: Cornell University Press.

Zehr, S. C. (1994) Accounting for the ozone hole: scientific representations of an anomaly and prior incorrect claims in public settings, *Sociological Quarterly* 35: 603–19.

Zuckerman, H. (1988) The sociology of science, in N. Smelser (ed.) *Handbook of Sociology*, London: Sage, 511–74.

Zürn, M. (1998) *Regieren jenseits des Nationalstaates, Globalisierung und Denationalisierung als Chance*, Frankfurt am Main: Suhrkamp.

Index